MEMORIES FROM A CULTURALLY INSIGNIFICANT WAR

By Joseph Murgia

Published by THC Publishing

MEMORIES FROM A CULTURALLY INSIGNIFICANT WAR
© 2024 By Joseph Murgia

For permission requests, please contact
dispatchestoauthor@icloud.com

Published by: THC Publishing

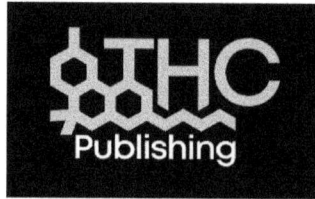

ISBN: 979-8-9918339-1-2

Printed in The United States of America

This is a non-fiction work. Names, characters, places, and incidents are either the product of the author's imagination or used fictitiously. Any resemblance to actual persons, living or dead, events, or locales is entirely coincidental.

For my family and my chosen family.

Contents

PREFACE

B ack in the early '90s, I liberated the country of Kuwait. You may have read about it in the newspaper, whatever that is. I didn't do it alone. I had a group of twenty-six thousand plucky co-workers with boots on the ground helping me out. According to the media, we completed it in record time, only six days, with no sleep and eyes full of horrors; it felt like it took longer.

It wasn't difficult for our Army. We expected more of a fight and pockets of resistance, but not something that would be difficult for the most powerful Army ever to exist. Considering they were fighting us with all of our old World War II and Vietnam-era equipment and weapons. I experienced a barrage of trauma daily. I almost died a few times. I saw and experienced things I wish I hadn't. I spent six months in Saudi Arabia and six days in a beige, sandy hell in Iraq. It was both the best and worst thing that has ever happened to me.

Here are some things you should never see, hear, or smell at nineteen or any age. I spent the last thirty years pushing these memories deep down. These memories used to haunt me daily; Now, after thirty years, they have mostly faded into the background. I saw truckloads of Iraqi troops burning alive, screaming, and begging for death. The terror in their screams is unlike anything I've heard before or since. I used to listen to their cries in my dreams. I saw a man bifurcated by fifty-caliber machine gun fire, his upper body flying end over end, his legs still standing on the ground. A crew of Iraqi men liquidated by gunfire turned into literal puddles of viscera.

I saw a human face ripped from a skull, lying in a puddle of magenta—an eyeball was hanging from the ceiling of a tank. Men who froze in fear in their last moments, face locked in a scream. Their hands were gripping pictures of loved ones. So many fun memories

Groups of tanks and soldiers are being strafed by an A-10 plane firing a Vulcan cannon, six thousand rounds a second, death from above. Human beings are reduced to a pink mist. The smell of burning human flesh, blood, and hair hangs in the air. You can taste the blood in the air on your tongue. The unholy, burnt human scent mixes with the smell of gunpowder and diesel. Breathing all of this in has probably taken years off my life. Your nose and lungs burn. The ground was littered with body parts, heads, and limbs ripped from their body by high-velocity rounds—bodies of men hiding behind a hill, all expired, bones protruding from their skin. I'm pretty sure I stepped on a gall bladder lying on the ground; it popped under my weight.

I watched an Iraqi tank drive over an anti-tank mine. The old, multi-ton tank launched into the air, tumbling end over end, with an orange flame that seemed like a feather. The tank tread flies off, resembling a ribbon dancing in the air. Men were thrown from the tank, rag-dolling, limbs breaking, bodies twisting. They finally come to rest, lying in a most unnatural position.

I heard on television that it was a video game war when I got home. Patriot missiles launched automatically at SCUD missiles, swatting them out of the air. That was the media narrative at the time. Being on the ground in a hostile country is different than controlling a missile from the safety of a ground control station thousands of miles away. Small arms were coming at us from every direction. The sound would echo in the valley, and it was very confusing where the rounds were coming from; anyone could hit you at any time.

This chemical cocktail of war messes you up, and you don't notice in real-time, but once you experience it, it never leaves. It digs deep into the psyche. It festers. I've never felt like myself again. You feed it with your fear; it grows like ivy climbing a wall.

Fear turns to anxiety. Anxiety turns to fight or flight. Panic sets in. You can't sleep. Once you eventually drift off, the ghosts visit you: black eyes, silence.

For years, I would wake up covered in sweat, heart pounding. I would then grab the stash bottle of booze I kept beneath my bed and take a big drink. I stare at the ceiling, waiting for the alcohol to numb me just enough so that I can drift off. It's horrible sleep, but it's still sleep. I repeated this for fifteen years until marijuana saved my life; it sounds like a joke, but it isn't.

War was the most intense drug I've ever experienced. Fentanyl as a life experience. No drug I have taken since has ever made me feel that alive. Our base selves are just trying to survive. Nothing more. Murder and fear of the unknown have taken humans very far in a relatively short period. War is about money, geographical power, and using poor people as cannon fodder. You don't get rich by not knowing where the bodies are buried. War is also fun; I was away from home for the first time. No parents telling you not to play with leaky dynamite. The war shaped me into who I am today, for better or worse. It pushed the introvert in me deep down and let the extrovert out. I survived, and I appreciate it every day. The damage to my psyche was also severe; it would take years for the fractures to be fully seen. Death imprinted on me during my tour of duty in Iraq. I've always felt like I was on borrowed time, probably just the survivor's guilt fucking with me, but I think it all the same.

This is a coming-of-age tale of discovering who I am during combat and entering the real world. I was going to a place where no one knew me; I could reinvent myself. I was an introvert, tired of keeping my thoughts and feelings inside. I wanted to let them out, and I could die sooner rather than later. It stuck, and I was no longer that introvert. I spoke up for myself and didn't put up with things I usually would. I successfully changed my personality and overcame that fear. Facing death makes you grow up very quickly. I experienced so many "firsts" during my military deployment. My first fake ID. My first time being shot at. This is the first time I've been drunk. For the first time, I had the genuine thought that I might die.

Some of my firsts were better and more enjoyable than others; the less enjoyable ones contributed to my future PTSD, which I immediately began self-medicating the second I got back to Fort Stewart, Georgia, my station, for the remainder of my time after Iraq.

I spent years self-medicating with whatever I could get my hands on. Drinking helped quiet the nightmares, not enough, but just enough. Alcohol offered a temporary escape from my meat cage. I didn't want to be haunted by ghosts anymore; I just wanted a dreamless sleep. More than thirty years after the war, I remember how war made me feel—a one-time, six-day bender filled with anxiety, dopamine, fear, fun, and facing your mortality—four days of no sleep breaks. Possibly a mini-psychosis, self-diagnosed. At the time, we thought the war would last months, not days, and it was a real fear that I could be fighting in the desert for at least a year. Luckily for all of us, this did not come to pass. I came home, served the rest of my time, and finally made it home. I thought I was done with the military, but there was no escaping the trauma I had experienced. It began to manifest in strange ways. I had no idea what was happening. It worsened every year, with new issues emerging. It took me almost twenty years to seek help.

I know the experience has profoundly and permanently affected me. It was stupid for Iraq to start a fight with the most powerful military on Earth. Never poke the sleeping bear. I blame that ass-clown Saddam Hussein.

I turned fifty as I began this memoir. They say you are replaced on a cellular level every seven years; I feel it. I was a different person at nineteen. I don't even understand that person when I think back. I'm so old. I remember when pronouns were basic grammar rules to help sentences seem less repetitive. No one had any issues with them until a few years ago. We live in an irrational world.

New experiences and world events outside your control shape who you become. My idealism juice is running low at this point in my life, but back in the 90s, my love for my country was at an all-time irrational high. So was my naivety.

This was before the Internet, Tarantino, Nirvana, cell phones, DVDs, and the events of 9/11. There was no social media, so I used a disposable camera, like a caveman, to take photos. If you still have an AOL email account, I'm judging you internally; please be aware of this.

The 90s were the last great American decade, and maybe I miss my youth. America seemed to be on an upward trajectory. Flannel shirts cost $80, thanks to grunge shifting the culture a few months after I returned from Desert Storm. America loves to fleece its citizenry with pop culture merchandise no one needs. It was a time when people protected their cars with the club. We had weathered the excesses of the 80s and were experiencing significant economic growth. The emergence of the internet and all the discoveries that came with it marked the beginning. We no longer needed the Thomas Guide (a book of maps in a grid so you can find your way around, pre-GPS) or a sextant to get around; being able to print out directions seemed like sorcery.

The first time I saw an email, I thought it was a magic trick. Automatic teller machines were new to the world. Everything used to involve paper; now, I only use paper to wipe down surfaces. I miss feeling hopeful about the future. I'm not even hope adjacent nowadays.

The stock market was booming, unemployment was low, and gas was cheap. Then, of course, we had Oklahoma City and the bombing of the World Trade Center, but these events did not diminish our hope for the future.

People, in general, don't even remember the World Trade Center bombing. The LA riots, Rodney King, and the white Bronco chase down the 405 swallowed the headlines back in the day. In contrast, the 9/11 attacks would shatter our sense of security and change everything. Some aspects have improved, but others remain unchanged. I miss picking a movie based solely on the cover art. I miss wandering around the video store or Tower Records. I miss double features and Tuesday night movies for a dollar.

I'll tell you what I don't miss about the nineties: rewinding videotapes before returning them to the store. The phrase "Be kind, rewind" holds little meaning nowadays. I don't miss phone books or calling 444-FILM for movie times. I don't miss the sound of typewriters or noise pollution. I no longer have to wait a week for my film to be developed.

I do not miss wired remotes and corded phones; I spent years untangling wires. I'm glad people can no longer smoke indoors. Snail mail can eat a dick, and so can carbon paper and whiteout. I don't miss having a book of phone numbers and dialing 411 for everything.

More than anything, I miss your boss's inability to contact you at all hours. Once you leave work, you take your phone off the hook; just being unreachable is now a gift. It used to be the norm. This needs to come back. The nineties were more straightforward, with an excellent economy, no division, and fewer responsibilities.

The nineties are inextricably tied to my youth; I never looked or felt better, so we are all nostalgic for the life I felt in my twenties. I had abs; those were the best three weeks of my life. Most people are at the peak of their beauty during this period. My mind was a steel trap back then, but it's no longer as tight as it used to be.

Catching up with former war buddies, we recounted the same events but in a different order, yet others in the same way. Days, like memories, bled together for me and my buddies. They reminded me of forgotten things, so I am grateful for their input. It's incredible how just a tiny reminder can bring back a flood of memories.

I've purposely left things out because I'm unsure if they happened. My friends could not confirm these events, so they will not be added to this memoir. This is my recollection of the linear timeline from basic training to flying home. I was operating on no sleep and running on pure caffeine and horror.

I remember some real-time conversations in detail, but I am paraphrasing all these years later. I was an on-the-ground witness to these historical events—an eyewitness to history. I've witnessed history before, but I've just never been a part of it. Desert Shield/Storm is a forgotten, culturally insignificant war, but not by me or the people who fought it.

Trigger warning and also a spoiler alert: PTSD causes suicidal thoughts; it did for me, but your PTSD may differ. I tried to be accurate regarding our language in the 90s. Inappropriate words flew out of our mouths with reckless abandon, no one even considering how they made anyone feel.

When I read back what I wrote, I clutch my pearls, so be warned. You may be triggered by our nineties indifference. One last thing: war is violent, to state the obvious. If you made it through the first few paragraphs, you'll make it through the rest. I describe the carnage as I saw it. I wish I had never seen it, but here we are. This is my After-Action Report (AAR), arriving thirty-plus years late.

CHAPTER 1

I was born in the 1900s; you heard that right! Raised in the last century. I'm a 20th Century Baby. As far back as I can remember, I've always wanted to be a soldier. I really can't tell you why. I had star-spangled eyes. No one in my family served unless you are talking about my mother's side, and I believe they fought for the Confederacy. Not a good track record there. They were traitors, and I'm embarrassed they are a part of my lineage. I had uncles and cousins who served in World War II and Vietnam, but I was unaware of their experiences until I researched this book. I power-watched war films growing up. You might think this would deter me from wanting to serve, but it didn't. I grew up in suburbia, and no one can hear you scream. So, that might be part of it: a desire and need for adventure.

I viewed the military as an exciting challenge. Travel the world, meet new people, and then kill them. I don't have any bloodlust, except when driving in traffic for extended periods. My childhood has a few interesting stories here and there. I spent much of the '80s avoiding the Noid and listening to the early rap stylings of MC Scat Cat. But there isn't much to say. Spoiler alert: I survived childhood.

This is the memoir of nobody, a random soldier, one among many. I don't mean this in a derogatory way. I'm not famous. I haven't cured any disease and never found a significant company or organization that changed the world. I never ran for office. My cult, Joeism, never took off. I put together a pamphlet, but no one was buying it. I'm not that deep either; I've stepped in deeper puddles. No real wisdom is to be found other than this: If you are struggling with mental health, you are not alone. We are legion. There are billions, and there is help.

Here is a quick summation of my uneventful childhood in broad strokes. Born in New Jersey in the early seventies, my family moved to San Diego, California, when I was a husky-panted six-year-old. I am of Italian Irish descent, with some French, Dutch, and German mixed in for good measure. I also have a small percentage of Neanderthal; you can see that in my Sasquatchian gait. If you could look at my DNA under a microscope, it would have a distinct limp; I'm sure of it. I can't believe I came from a group of hunter-gatherers unless they hunt and gather diabetes. I'm one generation off the boat; my father came to this Country from Italy when he was very young.

A little about myself: My love language is sarcasm. I've always felt that I might be on the autism spectrum. I read somewhere that most people are. I've never been able to separate English muffins properly; that must be a rung on the spectrum. I feel autismal, but I'm probably high-functioning. I'm not sure that is a word, but it is now.

I'm high-functioning for sure, but I've always felt disconnected from the rest of humanity. This is perhaps a very normal feeling.

I have an unnatural hatred of mimes and barbershop quartets. Suppose there is a barbershop quartet convention or a gathering place where they meet regularly. It should be carpet-bombed; alright, that might be a bit much. The only thing I hate more is California traffic.

Living in San Diego, I have a diet centered on burritos. I'm about forty percent carne asada. I identify as muscular, but I currently have a cadaver's physique at fifty. I'm trans-muscular.

I remember everything being orange in the seventies. Our carpet was orange, and the tile was a shade off orange. Typical seventies parents only interacted with us when we were in trouble.

I grew up fifteen minutes from Miramar, Top Gun School, so jets and the Blue Angels practicing in the sky above us were everyday occurrences. The late seventies and '80s couldn't be further from today. A not-so-glorious time when everyone played a game called "Smear the Queer." This was played at home and at school. Teachers would shout and point at the kid who was now the queer. "Jerry's the queer! Annihilate him!" What a time to be alive.

AIDS and Aquanet broadly defined the 1980s. Phil Collins sang every fourth song on the radio. We had toys that could kill you. Toy manufacturers were sadists. I would say toys killed 75 million kids in that decade without looking it up.

My parents were raised on the East Coast. All the stuff we had back in the day fell off a truck, is how I heard it. My dad hung out with unsavory people. I don't remember much, but I remember the cold when walking to preschool with my mom. When I moved to San Diego, the cold became a distant memory. People wear cold-weather parkas if it gets under sixty in San Diego. It's fucking ridiculous.

The plan was to stay a year in California and return to New Jersey, but you will never want to leave once you experience that perfect weather. You pay the sun tax to live here: low paychecks and a high cost of living. Luckily, my father made a decent living as a machinist. I never knew how much he made; that was kept from my brother and me. There is no snow to shovel in winter. He bought a house and planted roots.

My brother was born three years after me. He was my punching bag and test dummy for most of my childhood. My parents were checked out. When I was born, and they took me home from the hospital, they probably threw me in the trunk. It was a different era, with minimal safety regulations.

Home life was a typical upbringing for Boomer parents. I had to ask for affection; it was never offered outright. I'm not sure what fucked me up more, the war or having no affection growing up. I could not speak to my parents. They didn't seem interested. My parents were just hands-off. Mom read a book in another room. No one interacted with each other; we all had our interests, and we were left to pursue them.

I escaped with magazines, movies, and video games. I've never seen a healthy, loving relationship in my life. I saw beautiful relationships outside my family, but I spent years thinking that yelling at each other was normal. There was always tension in the air. My parents did not appear to have a love language. Their love language was indifference. It was a volatile household, and voices were always raised.

I was raised without religion—one of the best things my parents did for me. I never had to overcome indoctrination. I've read the Bible; Dan Brown should be hired to revise that book; it needs it.

My mom barely spoke until she was twenty-seven, and she couldn't drive until she was twenty-three. She was a stay-at-home mom for a few years until we were old enough to care for ourselves. Then, she transitioned into a career as a paralegal. She remained a paralegal until she retired.

My father and I had a complicated relationship. A story as old as time. Generational bullshit. We shared the same name. I had a middle name, but he did not. Most sons and fathers have the same familiar account. My father was physically there but not emotionally present.

Once he got home, his default setting was *fuck off*. He became one with his Barcalounger. He plopped down in front of the television, ate his dinner, and said nothing unless it was to criticize how I spent my time. He spent his weekends underneath our cars, with a cigar in his mouth, fixing the vehicle and avoiding his family. I never saw him exercise or breathe any clean air. My father was six feet two, bald. He was Chewbaccan, very hairy. It was not fur but a pelt.

You would need a license to hunt my father. I was not like him. I wasn't mechanically inclined, and he didn't seem to have any patience to teach me. I can change the oil on a car and a tire. That's about it. My dad could fix or build anything.

My dad's aggravated tone whenever he spoke to me constantly raised his voice, never calm. This kept me on edge my entire childhood, and I never knew what would set him off. He had worked since he was nine years old.

Pre-child labor laws, you could send a post-toddler, pre-adult kid down a hole into a mine with his Honeymooners lunch box and a pack of Lucky Strikes. My father was born in 1938, a Depression-era baby. He was an immigrant who got off the boat and didn't speak English until he was nine. We weren't born until his late thirties; my mom was nineteen.

My grandfather's father was a monster. I always try to think about what my dad went through. He was given coal for Christmas, for real, lumps of coal. They also fed his pet rabbit to him and told him afterward that they did not have much food.

My dad helped his dad build their first home and their first car. My dad was constructed differently. He used to turn the bacon in the pan with his fingers. I'm unsure if he felt pain; I did inherit a high pain tolerance. Being from Jersey, I heard many racist things come out of his mouth, some very creative, others not so much.

He never hit me, but I could tell he was thinking about it some-times. I'm sure I was a nightmare. All kids are, but I wanted attention and wasn't getting it at home. He would constantly make fun of me, not in a joking way. He would try to make me feel bad. I was nothing but an idiot or dummy—never any positive reinforcement. If I got a raise, I wouldn't hear congratulations but something like, "You know, that just means you have to pay more taxes."

This would make me rebel. *I'll show you a failure.* I would not do assignments. I only hurt myself. I would do the opposite of what he wanted me to do. I reflect on how childish this reaction was. *Why try?*

I could never do anything right. This was my stupid way of thinking then; the only person I hurt was myself, self-inflicted stupidity.

I would try, and I would be called stupid or useless. After hearing it for years, you begin to believe it. I attempt to remember the good things. I have forgiven him many times, but the anger remains; I'm not sure it will ever disappear.

I wish I could let it go. I try, I do. I focused on the positive things he did for me. He took me to the movies weekly; movies were my escape from the monotony of suburban life. He would take us to anything; he didn't care what it was rated. I saw so many great and awful films with him. The rating didn't matter. I saw every Chuck Norris and Van Damme masterpiece in the theater. He took me to horror films when my mom wouldn't take us. These movies profoundly shaped my life and introduced me to the world beyond my small neighborhood. I never wondered where my next meal would come from, so I thank him for that stability.

I was lucky to have made friends in school who had terrific parents. This made me self-reliant and incredibly lonely at the same time.

They would teach me things I wasn't getting at home. The first time my friend's mother watched me cut a steak at a family dinner, she looked at me as if I were a chimp learning to use tools for the first time.

A few incidents shaped me during my childhood. A medical issue affected me the most as a child: I suddenly lost my hearing. It pushed my life back one year. I would have graduated in 1989 and started my military career a year earlier. I might not have had the same group of friends in high school. My hearing loss baffled the doctors. Perhaps it was a virus, but we could never determine the cause.

My fourth-grade teacher, Mrs. Smith, thought I was lying and put me in the back of the class to "teach me a lesson." I missed the entire year because I could not hear anything she said. All I could see was her mouth moving. Because of her cruel indifference, I ended up staying back a grade. She could have helped me, but she chose not to.

A few months later, I woke up and could hear again, just like that. It took a little longer to regain strength, but I fully recovered. I feel fortunate to have stayed back a year, which allowed me to meet my friends who would be with me until the end of high school. I met my clique. The years passed, and I barely made it through middle school.

I started high school in 1986. I wouldn't say I liked school in general. It is state-sanctioned babysitting, allowing parents to go to work and keep the economy rolling. I wasn't dumb, but I was bored. I gained more from being with my friends on break than from any time I spent in the classroom.

I got my first job at McDonald's at the age of sixteen. This job was essentially assigned to you in suburbia. I only worked there twice in six months. They hired thirty other people the same day they hired me. They only called me in when no one else was available. Technically, I still work there; I never let go or quit.

I take that back. A few years after I left the Army, I received a postcard stating that if I brought in my polyester McDonald's pants, I would receive a coupon for a free hamburger. Now that I think of it, I was fired.

I took the ASVAB (Armed Services Vocational Aptitude Battery) test during my sophomore year. I did well enough to get into the military, but that is not saying much.

My family's financial situation significantly influenced my decision to consider a career in the military. The G.I. Bill was attractive, providing a thirty-thousand-dollar bonus after four years of service. I thought it sounded like a good deal. My grades could have been better. I had a few options. I could join the Army or attend the Barbizon School of Modeling.

Jokes aside, I visited different recruiting places around town during high school. I spent a weekend with the Navy SEALs in Coronado. To say it looked difficult is an understatement. I thought the Air Force seemed too easy. That is how dumb I was. Marines were a no-go since, living in San Diego, they surrounded you. We were just forty-five minutes from Camp Pendleton. I knew I would need to be a better fit. I visited an Army recruiter in Poway, California.

They use the ASVAB test to determine your skill set, where you might fit in, and your job—MOS, in military speak. MOS stands for *military occupational specialties*. The recruiter reviewed various jobs for which I was a good fit.

He kept returning to Combat Engineer, MOS, 12 Bravo. I might've been too dumb to do anything else. He explained that they are not merely grunts but also do construction and demolition work. Blowing things up caught my eye. He made it sound like I would wake up, blow something up, eat lunch, blow more shit up, and then have a healthy dinner.

He was speaking to my inner Oppenheimer. Combat engineers, or sappers, build bridges, but the kicker was that I would get to work with land mines and explosives. How could I resist? I signed up for the delayed entry program at seventeen, which would knock a year off my commitment; I would only have to be in uniform for three years instead of the typical four. My last year of high school was considered a year in the military. I was sworn in on July 11, 1990.

I graduated with no honors—very little pomp and circumstance. That was the first time I realized how great it was to be a kid. I wanted to grow up so badly. I discovered I would never have a summer vacation again. It was a gut-punch realization. It was a bit of existential dread. I had two months of hanging out with my friends, going to the movies, and stress-eating before taking off to Missouri. You never know how good you have it until it's gone.

CHAPTER 2

July 11th, 1990, was the first day of the rest of my life. The only other time I had been away from home was during eighth-grade camp, where I made a shitty napkin holder and a disfigured candle that I was proud of. But this was different; this was the first night of the rest of my life. I was excited, scared, and unsure of what the future held. Ultimately, I had the opportunity to make my own mistakes. It was a long day, starting at 4:45 a.m. when my recruiter picked me up. We picked up a few other people on the way. We had a few quick meetings but were left to our devices. We were flying out at 6 a.m. the following day. My mom picked me up, and we had lunch. I napped, and she took me back to the hotel at 8 p.m. As we pulled up, I spotted a girl sitting in the doorway, smoking a cigarette. She was pretty. I said goodbye to my mom again and headed towards my room.

The girl makes the first move and starts a conversation before my key hits the door. I'm glad she did because I had no idea how to start talking to her. I wanted to, but I was too shy. She starts slinking towards me. She makes small talk and asks me to come to her room. She wants someone to talk to. She also has alcohol.

We sit and talk; there is something in the air. She is sitting close to me. I stare at her beautiful face, a symmetrical American. My favorite type of American. Her legs looked tan and smooth; she even had pretty feet. I'm not into feet at all; they are disgusting. I never noticed them, but this time I did. They are utility appendages, not sexual turn-ons. She smelled like vanilla and hope. As inexperienced with women as I was, even I picked up on the sexual energy. The kissing begins. I slowly undress her, and she excuses herself and goes into the bathroom.

When she emerged, she was wearing granny panties, like old-timey bloomers. She inexplicably had a bowler hat and tie on now. It was as if she were wearing the Annie Hall lingerie collection. This is everything I thought being an adult would be. My first mouth hug exceeded all expectations. I felt like I had crossed a significant milestone in adulthood. This is my life now.

I thought my time would be filled with alcohol and blow jobs. It was mostly alcohol, sadly. My focus in life was to meet women and touch their bathing suit areas with consent, of course, and vaginas from other great nations too. *All vaginas welcome* is my motto and always will be, quite frankly. I had no idea when that would reoccur; that experience excited me for the future and more mouth hugs.

We woke up early, hungover, and headed to the airport the following day. We said our goodbyes, and the next thing I knew, I was on a jet in coach heading to the middle of America: Fort Leonard Wood, Missouri. *Basic training, here I come.* We landed in St. Louis, which was in rough shape. The view from the bus was insane. It seemed like entire neighborhoods were burned-out husks of buildings. I made a mental note not to visit. We passed the gates of Leonard Wood.

We are all familiar with what basic training entails. You've seen a movie. My experience was like *Full Metal Jacket,* with some laughs. The first words I heard as I got off the bus were, "*Drop!* Give me twenty!"

I dropped into the pushup position. Why did I not train for this? My arms began to shake like a broken washing machine. This was on the first pushup, and I owed Sergeant Aneurysm nineteen more pushups, which didn't seem likely to happen.

Then, finally, he turned so red in the face while yelling at everyone that I thought he would die. I knew that was their job, but take it down several notches before you stroke out. From that point on, I would be *Private Murgia, fuck face, hey you, get over here,* or *retarded Karate Kid.* I suppose I had a passing resemblance to a special-needs Ralph Macchio. We were marched over to our temporary quarters before being assigned to a platoon. We put our suitcases in our lockers. We then marched over to the pay office. We were given our first military paycheck. Most of this check would be gone in a couple of hours. You bought your uniforms and boots; they were not given to you. They were not cheap either. You had to purchase two uniforms, two pairs of boots, and two pairs of physical training (PT) clothing. We were then allowed to call our loved ones. This was the end of day one.

The next day, we shuffled down to the barber. I do not have a head that is made to be shaved. You didn't know this until you shaved your head for the first time. It was a crap shoot. I watched as all this hair was shaved off, and you saw these monstrous skulls, never exposed to the light of day before, in some cases. Some people are lucky and have good-shaped heads. My head looked like Sloth from The

Goonies—not great. My head needs hair. I see actual deformities, not Elephant Man levels, with weirdly shaped skulls that should be covered by hair. A phrenologist would have a field day. I don't look good sans hair; now, I go to the barber and say, "Give me the Rachel Maddow!" and I'm on my way in fifteen minutes.

Next, we had our dog tags created. I gave them my name and social security number. They knew my blood type. They asked me about my religion. I said none.

"You're an atheist?" he asked.

"I guess I never labeled it before, so yeah, atheist," I replied.

"I can't put that on here."

"Why?" I asked.

"It's against my religion."

"It's against your religion to type the word *atheist*?"

"Yes."

"That makes no sense whatsoever. Can't you type a word? So, you can type something else that is a lie but can't type something that is the truth? So, your religion is good with you typing a lie?"

He ignored me. "Your last name is Spanish?"

"No, Italian, mostly Irish, though."

"Spanish, Irish, Italian, all the same shit to me. You're Roman Catholic; that's what I'll put."

"But I'm not. I don't believe in a god. What is this shit?"

His religion superseded my non-religion. I didn't know I could put my foot down and fight it, but they just wanted to move the line along. I felt like a kid. He was an adult in my eyes, so I didn't want to fight it. I was nervous enough.

Reality crept in, and I felt I had made a colossal mistake. I wasn't prepared physically, emotionally, or mentally. I guess that every person has this epiphany on day one. I had to adapt quickly.

I'd watched so many basic training scenes in movies that you'd think I'd have prepared better. This should not come as a surprise, but watching others get yelled at is often easier. It is something else when you receive loud noises and spittle on your person.

We were assigned a "battle buddy." You worked with this person. If they messed up, I got punished. I had no luck; my battle buddy had borderline special needs and could not do anything right. I paid the price.

We were then issued our equipment and had to try it on. The drill sergeant inspected each person, ensuring they correctly attached everything to the web belt. My buddy had everything upside down and backward. Pushups for me again.

We then began marching drills. We had marched before, but this was when we started to take it seriously. Essentially, we had to keep step with left, right, left right. You would think these were the first time any of us heard these words. We looked like a bunch of idiots in the beginning. That became my introduction to cadence. Cadence is a song the drill sings out to keep you in step with each other. We heard all the greatest hits. This did help you keep time. It is a skill that takes a bit of musicality to get it right. We also had drills to change the lyrics of the cadence.

Nothing moves fast in the military. You do this over and over, all day. Hurry up and wait—typical army. They would mostly be funny, sexual lyrics and done with no one around.

Went down to the schoolyard, where all the children play
I pulled out my Uzi and began to spray
I went down to the store where all the people shop
I pulled out my machete and went CHOP CHOP.

That's a fun one—- a lyrical murder spree.

Vaccine day was an early part of the process. We lined up in a large, white room with men on each side, standing next to a pneumatic vaccine distributor. The line seemed to go on for a few hundred yards. *How many shots am I getting?* It shot the vaccine into your arm using air, not a needle. They launched it into your body. I was expecting needles. You were told to keep your arms down by your side as you moved down the line. I got around fifteen vaccines all at once. Of course, this is impossible when something hurts, and I flinch, and the air gun slices my skin, and I begin to bleed down both arms as I move to each nurse.

My blood was free-flowing off my fingertips as I reached the end of the line. It was a heavy flow day for me. We all left a blood trail behind us. First, they gave me two small pieces of gauze for the extensive bleeding. Then, the drill sergeant started yelling at me because I did not follow instructions and moved my arms.

"Drop!" he yelled.

I couldn't lift myself off the ground, as usual. But, as I was trying, blood was now pooling below me. The drill sergeant saw that I could not lift myself off the deck, took mercy on me, and told me to get up, and the nurse taped me up. He was shaking his head at me as I walked off. *Sorry for getting blood all over the dirt in Missouri.*

I must've been a real disappointment. That was just day two.

I had been mainlining sugary soda since I was a kid. There is no soda or coffee in basic training—just juice. If you have any caffeine addiction, you have to go cold turkey, and sugar is a hell of a drug to ween off while doing more exercise than you have done in the last two years.

The headache starts small but then begins to spread around the skull. The answer to every medical issue in the military is Motrin (ibuprofen). Broken leg? Motrin. Bullet wound? Motrin. They would not give me Motrin for my headache. They told me to suck it up.

It took two weeks for the headaches to go away. I don't know what heroin withdrawal feels like, but I'm guessing it is in the same neighborhood as coming off sugar and caffeine. You pile brutal daily exercises on top of the withdrawal, and I'm surprised I didn't pass out daily.

Basic training is where you assimilate into Army culture. It is a dehumanization program that breaks you down and builds you back up. You might not like the version of yourself they represent.

You are trained not to look at the people you are fighting as individuals but to view them as enemies. They are viewed as other, essentially non-player characters, you put a bullet in. They are not people with families, dreams, and emotions. Thinking of them as people will make it harder to kill them.

My upbringing in paradise worked against me. I was a semi-pampered Californian who had never camped outdoors. I played Dungeons and Dragons, which exercised my imagination. My idea of roughing it was reading *Fangoria*, *Nintendo Power*, *MAD* magazine, and *Inside Kung-Fu* in the backyard. Al Jaffee and George Carlin raised me. I was also a connoisseur of *Ninja* magazine—yes, there was a Ninja magazine. I assumed I would be a ninja as an adult; I read it for the articles.

Did I know martial arts? Not at all. I took a few Kenpo classes and one Ninjitsu class. They just beat me in the Ninjitsu class. I learned a little. All those years of fantasizing about being a soldier were destroyed in just a few days of reality. Maybe college wasn't a terrible option. California is filled with community colleges that don't care that your grades make you seem like a student with special needs.

On the third day of basic training, they lined us up alphabetically outside the drill sergeant's office. My bunkmate, also named Joe, was in front of me. He also felt he had made a mistake and wanted to go home. They called in each soldier individually and asked them if they wanted to be there. Everyone in front of us went in and came out, all stating they wanted to be there. This was voluntary, so this was your chance to get out.

My bunkmate, Joe, went in, and they asked him if he wanted to be there.

"No, I don't want to be here," he said.

They closed the door, and I heard every drill screaming at him, calling him names. He told me later that he was slapped a few times. I did not witness that, but they couldn't touch it all.

I was about to go in and say the same thing, but I heard him reverse course through the door. I went in and told them I wanted to be there, but that was a lie. My main antagonist, the drill sergeant, was named Sgt. Himmler—not his real name, but the same temperament. He was built like a Lane Bryant mannequin. Sergeant's mustache. They all wore the Army-issued brown Stetson.

He would give an order, and I would question it. I couldn't help it. It was a sickness. Every authority figure was my dad. *Why are we doing this?* Very stupid. He didn't like me and looked me up and down with disgust. He would seek me out to torture me, having me do pushups on gravel. I still have the scars on my hands. The other drill sergeants were different. It was just a job for most. We had one drill who would constantly brag about the woman he was with the night before.

"You boys want to smell my fingers?" he would taunt us daily.

I was sheltered and baffled as to why anyone would want to smell his fingers. That was just gross. I was so naïve and completely oblivious. I remember other guys around me saying the same thing: we were all inexperienced kids. We are expected to kill. He was constantly talking about fingering ladies; I like to give a lady a good thumbing like you are rubbing out a spot. He had a funny lisp, but I did not dare make fun of it. If you had a girlfriend or wife in basic training, you were constantly taunted by her spending time with Jody, a fictional character created for psychological torture. He was fucking your lady the second you were out of sight, and there was nothing you could do about it. *Your wife or girlfriend has no loyalty to you. Jody will replace you. He will move into your house. He will be driving your car.*

It's some silly shit. Some guys would get so angry at the implication. It wasn't real; it was a mind game. You would think they would pick a more masculine name. I have never been or will be sexually threatened by a man named Jody.

July in Missouri is Chernobyl hot. This might be an understatement. It's the heat where it feels like your internal organs are also sweating. I could sense the blood slowing in my veins. It felt like my heart was pumping ketchup. The military has a heat index—anything over ninety degrees—and we sat in the shade.

I spent most of my basic training in the shade, learning little. Since they wanted an early start, we would start our day at 2:30 a.m. or 3:30 a.m. Shaving was mandatory—something I had never done before joining the military.

My face was follically barren, with only some light peach fuzz. I could escape not shaving, but I was wrong. In some cases, they would examine your face, nose to nose. Then, I was sent back to the barracks to shave. Sometimes, the drills would sit and watch you.

"I don't care if you see nothing on your face, shave!" the drill shouted. "You lather your face and drag a razor across until I am satisfied. You don't question what I say."

You had five minutes to eat your food in the mess hall. The drills timed you, and when you hit five minutes, they chased you out of the mess hall to make way for another soldier to eat. Anything you didn't eat got thrown away. Luckily, I ate like a duck. I barely tasted the food. You also had to guard your food, like in prison. Asshole drills would slap a tray out of your hand; no meal for you if that happened.

Every day was different, featuring a mix of exercises and classroom work. If you weren't training or in the classroom, you were expected to be studying The Soldier's Blue Book. It covered all the Army's rules and what was expected of you. It showed you how to set up your uniform, and you were constantly quizzed. It also covered how to march correctly. Have you ever wondered what to do if you are ambushed? It's in the book. They wanted the book memorized.

During basic training, lunch was typically an MRE (Meals Ready to Eat), but rarely a hot meal, followed by more training and a hot dinner. After dinner was the most crucial part of the day: mail call. It was a bad day when your name wasn't called. It was your only link to the outside world.

We had no television and were not allowed to make phone calls home. You then get personal time, and then guard. The barracks were solid brick, so I wasn't sure what would catch fire.

The only respite from training was pretending to be religious and attending church on Sundays. It was the only time I had ever walked into a church service. I love old churches when I travel, mainly for the architecture. Did the holy water boil when I entered the church? No idea. I didn't check. This was a way to rest my eyes and get some sleep. I had no idea how annoying the whole standing-and-kneeling-multiple-times thing would interrupt my rest. I went to Catholic services because of my incorrectly labeled dog tags, which ended up helping me out when they questioned my religiosity. My dog tags did the lying for me. I found a loophole. They had to let me go. I would escape for an hour or so.

On Sunday, we did light PT and then cleaned the barracks. It was usually a light day. Your bed had to be made every morning, but they would literally bounce a quarter on it and check the corners more closely on Sunday. Clean your locker. This was when most soldiers tried to find blind spots and catch a few Zs. Some would close themselves in their lockers.

One Sunday, a soldier did just that. The drill instructor locked him in, and they dragged the locker to the stairwell and threw it down the stairs with the kid inside. We were all kids getting physically abused by these bullies. The kid was hurt, and the drill knew he had gone too far. He started to try to cover it up immediately. I have no idea what happened to that kid.

CHAPTER 3

B asic training can be super fun at times. I loved hitting people with sticks. I had way too much pent-up sexual energy plus repressed familial rage, and I was leveling people. I don't know what came over me. A pugil stick is a long, padded stick on both ends, but it still hurts if you have enough force behind the hit. It was like adult pillow-fighting. We were placed in a makeshift Thunderdome and beat each other for several afternoons.

Going to the range and shooting my M16 was incredibly enjoyable. I learned to strip it down, clean it, and love it. I was skilled at disassembling and reassembling the rifle quickly. It was like a deadly puzzle. I never thought about maintaining a rifle's readiness. I always thought they were self-cleaning. I was an idiot, but I learned fast, and it was something I excelled at. My weapon was always sparkling. I could field strip it and put it back together in a minute and a half.

During the third week of training, I was minding my own business when this Black soldier I had met earlier in the week asked me the origin of my last name. I told him it was Italian and thought nothing of it.

Then, I saw him walking toward me briskly and holding a rubber training rifle. These are designed to resemble M16s but are made of solid rubber and are heavy. He had no expression on his face.

I thought he was coming over to talk to me or something. I began to speak. Just then, he struck me across my face with the rubber rifle and kept walking, never breaking stride. The hit dropped me to my knees. The pain in my jaw lit up all the nerves in my face. I saw a bright white flash, but it never went out. My vision was closing in on a pinpoint.

I forced myself not to go out; I dropped to my knees. My friend ran over and asked if I was all right. Of course, I was, but I was in a lot of pain and was primarily confused. What did I do? I never did anything but talk with that person, and it seemed fine. I had never been in a fight, not really.

I was in a one-sided fight where I was the only person punching, and the other person never stopped walking, so I just stopped. It wasn't a fight if it only involved three punches, and he never hit back. A black girl in my school slapped the taste out of my mouth once for no reason. That was my total history of violence.

My friend said I needed to respond, or that was just the beginning. They would never stop coming for me. I was not that type of person. I didn't want to fight. I was scared and didn't know what to do. My friend in the bunk above me said he would get my attacker to relieve me on watch; he was responsible for the schedule.

I needed to do something. I couldn't sleep. I didn't know what to do, but I no longer wanted to be a victim. We were all issued a rubber training rifle for the watch. My bunkmate's only advice was to get mad, see red, and hit that motherfucker hard. Send a message-prison rules.

My sentry post was in the stairwell. I thought about the pain in my jaw. I was trying to get angry, but it wasn't working until I heard the door open at the bottom of the stairs, and I saw the motherfucker who hit me. That was when I felt my blood begin to boil.

He greeted me with, "Hey, pussy."

I said nothing.

"I'm your relief, pussy," he said. "Hand me the rifle."

I slowly descended the stairs, and he began to stride towards me. I didn't know what to do. I was going to pick my moment. His whole-body posture was anticipating me, so whatever I needed to do had to happen fast. I had the high ground, Jedi-style. He got closer and reached out for the rubber rifle.

Instead of doing what he did, I acted as if I were handing him the rifle. I kicked the inside of his knee, the exact spot we were taught in hand-to-hand combat class the week before, and his knee buckled. He tried to catch himself by grabbing the railing, making the fall awkward.

He plummeted down the stairs, and his back and head slammed into the door behind him. The sound echoed through the stairwell. He rolled over to his side and grabbed his knee. I lifted the heavy rubber rifle over my head and slammed it down on his body as hard as possible.

I missed him, and the fake barrel hit him, but most of the rifle bounced off the floor. Mission not accomplished. I wanted to hurt him. I did some damage, but I was seeing red.

I opened the door, headed back to my bunk, and waited. I waited for him or his friends. No one came, and I stayed up the entire night.

He never bothered me again. He was injured, but not too severely. I saw him running with a bandage on his knee a few days later. We locked eyes. I flipped him off. The funny thing was that he and his friends turned their attention to my bunkmate, who made the schedule. He fought them off with some regularity. The amount of testosterone in the air was palpable.

It took me a week to let my guard down, but I was always ready for a rematch. I wasn't tough, but I got hit in the face and stood right back up. It hurt, but I managed to get through it. I learned an important lesson: I got punched in the face, but I could take it. A lesson every man should learn.

The drills constantly brought up the gas chamber. Everyone had to go through it; they wanted it to become muscle memory. The second you picked up that gas, you knew how to hold your breath and put that gas mask on.

We marched over to the gas chamber. It was a large, sealed chamber with a large window, allowing the drills to observe. We put on our gas masks and ensured a good seal. We were instructed to line up. They would release the gas, and then we would remove our gas masks and breathe. *Do not hold your breath.* We would not be let out if they saw anyone holding their breath. If one person did, then all suffered the consequences.

My group agreed to open our eyes and get it over with. We had to be out as quickly as possible. We entered the chamber, lined up, and filled the room with gas. We heard our instructors tell us to remove the masks. We all do, but one of our group's soldiers kept his eyes closed and held his breath.

He began to panic. We started to yell. My anger at this assclown superseded the gas for about one second. My eyes burned, and snot seemed to pour from every hole in my head, somehow out of my ears. The guy was in full panic mode and began to run around the chamber, still holding his breath like a chicken without a head. It was now getting intolerable. We rushed him.

Once you get gassed, you never forget it.

After what seemed like a lifetime, he opened his eyes and got a lungful of CS gas. We were coughing and struggling to catch our breath. At that moment, the door opened, and we stumbled onto the grass and fresh air. The first clean breath of air was nirvana. I used my canteen water to wash off my face and rinse my eyes.

Nothing seemed to work. I would dry heave with a hint of the gas smell, so I had to get far from the gas chamber. We stared down at the panicked soldier and shunned him appropriately.

Even a hint of that smell still triggers my fight-or-flight response.

We continually went back to the M16 range. I always looked forward to it. The first targets you hit were close, and they were stationary. I spent my youth playing Nintendo's *Duck Hunt,* so my hand-eye coordination was on point. I just imagined they were ducks, murderous ducks. The targets would slowly move farther away, eventually leading us to a computerized range with targets that moved up and down, much like a game of murderous whack-a-mole. A tower overlooking the range conducted a drill and several exercises after we finished firing our weapons. I remember lining up and spending half an hour "policing the brass." I love the smell of gunpowder on the range.

One day, after the morning range, we were all sitting and enjoying lunch when we heard a nearby gunshot. Everyone dove to the ground. Chaos erupted, and the drills ran soldier to soldier, examining them and getting them into the woods. Once the chaos had settled down, we discovered the source of the gunshot. One of my fellow soldiers shot himself in the foot, putting the rifle against his boot and pulling the trigger.

He stupidly thought this would get him out of basic training and on a plane back home. He was just ten feet away from where I was eating. He snuck the round off the range, loaded it in the weapon's chamber, put the barrel up against his foot, and pulled the trigger.

The Army did not send him home; they cycled him through the next class. We saw him a couple of weeks later in another company, running with a boot on his foot. He had to redo those weeks. Had he not shot himself, he would have been out of basic training sooner than later.

Next up was the grenade range. I wanted to blow things up. We were all looking forward to playing with grenades. It isn't working, but it felt like it was playing. The explosion is teeth-rattling, even with a massive cement wall in front of you. It shakes the marrow in your bones.

We also had to run the confidence course, jumping over walls, swinging on the jungle gym, and crawling under barbed wire in the mud while they shot live rounds over our heads. That was fun as hell. Then, there was the hundred-mile road march over four days. We marched twenty miles per day up a steep incline for several miles. I weighed 140 pounds at the time. We were required to carry ninety pounds in our backpacks. If your equipment weighed less, they would add ten-pound plates until you had ninety pounds. I had never been through so much hell, and we marched in spurts because of the heat index.

My legs and feet had never experienced pain and swelling, and my toes wanted relief. You couldn't remove your boots. Otherwise, your feet would swell, and you could never put on the airborne boots again. My socks would be red with blood. That was a similar experience to several Disneyland death marches I have been on. We also learned how to Bivouac, which is to set up camp in the woods.

You built a lean-to, a temporary shelter made of sticks. They told us to dig a foxhole with our entrenching tool. Missouri doesn't seem to have any loose sand anywhere in the state; it is all rock with a sprinkling of dirt. I tried to make a hole but only made sparks, slamming my entrenching tool into the solid rock.

Most of the Bivouac was spent sitting in the shade. We entered the nuclear winter phase of summer in Missouri in late August. The shade barely helped. About four weeks in, I injured my knee while running. It swelled up to the size of a grapefruit. No sympathy from the drills. Swallow a Motrin and fuck off. I was limping everywhere, and we ran harder and faster. Every night, the ache in my knee would make it difficult even to sleep.

This led to the Army Physical Fitness Test, which you did not want to fail. My knee was not doing me any favors, but I managed to pass. This was your way out of basic training and into the graduation phase, then heading to advanced training and, in my case, airborne school.

Going to airborne school was different, especially right out of basic training. There was a waiting list. I got lucky getting in right out of basic training. My drill instructor waited ten years before he was accepted. We were in the same class, and he had to take the same shitty bus ride I did.

Graduation day came. The drills moved down the line. They handed us our "diploma," shook our hands, and proceeded to the next soldier. I got mine. Sgt. Himmler shook my hand and his head. He tried his best to get me to wash out, but I managed to get through it. We were dismissed and off to blow off some steam.

I graduated from basic training weighing 138 pounds on my six-foot frame. I looked sick, way too thin. I was no longer formless yogurt but tighter-formed yogurt. Fort Leonard Wood was where my advanced training would take place, mainly with the same crew of drill instructors. They were all twelve bravos (12B). This is called one-station unit training. Advanced training is where you learn the skills required for your job. In my case, explosives, razor wire, and digging holes, sometimes with a shape charge—an explosive that makes deep holes.

We headed into advanced training, and I got through unscathed. I graduated again. We also got our orders and found out where we would be stationed. I did not want Korea. I heard the STDs were rampant over there, and if you got one, they would take you to an island and hit your pus-filled dick with a mallet. That was the rumor, as ridiculous as it sounds.

I was going to be stationed at Fort Stewart, Georgia. However, first, I boarded a bus to the Airborne School in Fort Benning, Georgia. A twelve-hour bus ride is always an enjoyable experience. I could not wait. Jumping out of a plane is like the ultimate roller coaster, and it is one of the main reasons I signed up in the first place. I wanted that freefall.

I'll never forget crossing onto the base, "What It Takes" by Aerosmith blaring in my headphones. We got there on a Friday. I saw the two-hundred-fifty-foot towers scattered around the base. My knee was aching, but I thought little of it. Adrenaline was taking over.

I loved the airborne training. It was fun and different from basic training. You were treated with respect. There was no rank; you removed it from your collar and soft cover (hat), so you were all the same: enlisted and officers co-mingled. It was also my first taste of Army life, where it became a six-to-five job with weekends off.

The first week was ground week. We practiced jumping out of a plane door, landing with our feet together, and doing a roll. Then, we move on to the swing landing trainer, where you jump off a platform thirty-four feet above the ground, hang from a suspended harness, keep your feet together, and slide down a wire. I could have done this all day.

My knee could not do that all day, and it was swollen. I finally got to ice it after running on it injured for weeks. I knew it would cause issues.

I felt a jolt shoot up my leg whenever I came down hard—airborne training involved jumping and landing on wood chips, which did not soften the fall. We were repeatedly jumping off boxes three feet off the ground. I also had severe pain in my right foot after jumping off a box. Great. It was something new to deal with. I tried to conceal my injuries so I could complete the course.

Airborne school made you run everywhere. You were not allowed to walk. You were punished if you were caught walking in a designated running area. I was running, and one of the drill instructors noticed my limp. He asked me if I was all right, and I lied. As we kept running, my knee buckled for the first time. I fell, tried to catch myself, twisted my knee, and heard a pop—pain shot through my leg. I could not put any pressure on my right leg. I begged them to let me go; I had one more week. The drill got me a wheelchair, and they took me to the medic and x-rayed my knee. I had torn many ligaments, and my airborne career was over like that—no wings.

I wanted those wings so badly. I was crushed. I was off to my permanent station at Fort Stewart, Georgia. There was no internet, so I couldn't research the base to see what I was getting into. I spoke to a captain at the airborne school who was stationed there.

I was told the base was empty because they had been shipped to Saudi Arabia. He was going to Iraq directly after airborne school. Thanks to the knee injury, I got to go home for two weeks before heading to Fort Stewart.

While I was in basic training, events began to unfold in the Middle East that would directly impact me and millions of people.

On August 2, 1990, erotic romance novelist and dictator—it's true, look it up—Saddam Hussein invaded Kuwait. My future unit, the 24th Infantry Division, was one of the first units deployed to Saudi Arabia. The operation was initiated with the participation of thirty-five other coalition countries and was named Operation Desert Storm.

My parents were distraught, but I took it all in stride. I didn't know how to feel, but worry would creep in. Am I going to die? This was the main thought; it would creep in, and I would push that fear to the back of my head. I boarded a plane to Savannah, Georgia, and headed to my unit. When I got to Fort Stewart, I was assigned a room in the barracks. I got in, put my luggage down, and hit the rack.

The note on my pillow read, "You're going to Iraq, and there is nothing you can do about it." Don't get me wrong, I like the direct approach, but that was the only note I've ever read that scared me. The reality was setting in.

I cried alone in my room that night. Part of me was also wildly excited. I just wanted to get going, no matter what the outcome. It was life or death, but let's get it over with was my attitude. I felt free.

As scared as I was, my parents weren't there, no matter where I looked—no one to tell me no. I had to make my own decisions, good or bad. I could indulge. I could indulge in whatever was needed to be gratifying. I was a naïve adult. I was finding my way. I wanted to get as much living in as possible. I want more mouth hugs, please.

Outside of Fort Stewart, to the south, was a small town called Hinesville, which was voted the most boring city in Georgia. You had to drive twenty-two miles to Savannah if you wanted any entertainment. That was the first time I had ever seen Walmart. My first time eating at Taco Bell was in Hinesville. San Diego had authentic Mexican food every hundred yards. The California Burrito is life, with carne asada and French fries, the ultimate meat and potato delivery system. Taco Bell was a cruel facsimile.

Georgia was like living on another planet; Everything was familiar but different. Alcoholism was the top hobby of most people living on the base and in town. The best part about this time was finding a military ID on the ground and turning it into a fake ID. The designer of the army ID stupidly put the picture on the front and the birthday on the back, so I glued his card to mine. I magically turned twenty-two years old overnight. I could now go to the bars in Hinesville and touch some ladies' bathing suit areas. They were teeming with women, primarily military wives and girlfriends. Their men were in Iraq, and they were out in force.

Monogamy is not popular on military bases. I spent two weeks going home with women to their military housing on the base. My nineteen-year-old penis did not care; it was like a divining rod for a drunken vagina. I'm not proud of it, but it happened. I pretended not to see the family photos on the wall as we went to the bedroom. I could not wait to disappoint every single one of those ladies sexually.

Planes were landing every week to shuttle people to Saudi Arabia. I delayed the inevitable for about two weeks. I missed two flights to Iraq by ensuring I was the last person in the queue for the plane. I would get right to the stairs and be turned around. It wasn't a war at this point, but it was heading in that direction. The planes were from a company called Tower Air. I had never heard of them. I didn't want to fly in some off-brand plane. It just made me more nervous.

My third attempt to be turned around by being last in line did not work. I had my headphones on, listening to the Beastie Boys' "Looking Down the Barrel of a Gun" from *Paul's Boutique*, which seemed a little on the nose as I climbed the stairs to the plane.

CHAPTER 4

A fter a twenty-one-hour flight, we arrived in Saudi Arabia in late September 1990. I was greeted with, "Welcome to Desert Storm."

Desert Storm sounds like an Arizona Jazz radio station. We were driven to a temporary base near where we landed. It was a tent city, a motor pool filled with Humvees, five-ton trucks, and deuce-and-a-half trucks (two-and-a-half-ton trucks). Antennas poking out everywhere. Crates of equipment are stacked. Soldiers unloaded storage containers of equipment—a tin-roof-covered mess hall with hundreds of tables. The weather was quite lovely. Dust kicked up everywhere I looked. I missed the hundred and ten degrees in the shade months my unit had suffered through. It was the dry heat I was used to; it felt like California in October.

I was assigned a bunk, a gas mask, and an M16 rifle. I had a week before I would be shipped off to my unit. For the next few days, we did some cultural training. When Saudi men speak to you, they get right up in your face, into your personal bubble. If you take a step back, this is an insult. We were instructed to turn our heads. We were warned about drinking alcohol. No *Playboy* magazines. It was no fun, essentially. Also, do not admire watches or anything; they will try to give them to you. Again, that was what we were told about Saudi culture. I'm not sure it's even true. We began gas mask drills daily. There was not much else to do, so I would sit in the tent city until I was told where to go, driving around a general in a Humvee or just listening to my Walkman. This went on for days.

We must zero our M16, dialing in the front sites and ensuring we can hit things with our rifles, shooting down range. My fellow soldiers and I headed to the range. The engineers built a few ranges, small arms, rifles, and bigger guns like the .50 caliber. The American presence and the extent of the base were impressive. America does not fuck around. We got back from the range, and I was greeted by a colonel who told me I was heading to my unit the next day, the 3rd Engineer Battalion. I could barely sleep that night. I loaded my rucksack and duffel bag and waited for the truck to arrive. A deuce and a half pulled up, and a guy with a Gollum-like physique jumped out of the driver's seat. Pencil-thin mustache, ragged uniform. His name was Wingood, West Virginian, and he was a good ole boy. He was the spitting image of the Crypt Keeper from *Tales from the Crypt*. I jumped in the front of the truck with him since the back was filled with boxes and other soldiers. His favorite actor? John Wilkes Booth. That guy was a dick.

He asked me where I was from. I said California.

"Oh shit, are you a faggot? You look as queer as a football bat. Jesus, how old are you? Are you a toddler? You have a baby face."

He keeps looking me up and down, trying to size me up. I probably didn't weigh any more than his anorexic ass.

"Yes, I'm a six-foot toddler, and no, that is not how sexuality works," I replied. "It isn't location-based. I hear the same fucking joke every time I tell a mouth-breather that I'm from California. Is the mess serving retard sandwiches? You look like you crawled out of a grave. Should I assume you fuck dead people? Where the fuck are you from?"

"West Virginia," he sheepishly said, like he was embarrassed. He should be. It's a beautiful but ignorant state. We crawl into the truck's front seat; he turns the key, and the engine roars to life.

"I knew it. I don't even know why I asked. You look inbred. Are you the leader of the hill people? Your sex ed class is probably called 'Sister Fucker 101' if you went to school at all."

We drove to a main road. For miles in any direction, there was nothing, just brown desert. The wind blew sand everywhere. This is the ghost of Christmas's future. This constant wall of sand blown up by the wind will be an hourly irritant.

"Jesus! Fuck you, motherfucker," he said. "Let's play the silence game for the rest of the way. We have a long drive. I don't even have a sister. I will beat your ass."

That was my first interaction with anyone in my unit. You could not let these people smell fear. If you did not stand up for yourself, they would eat you alive. It's a prison mentality.

"Silence is impossible," I continued. "You? Beat me up? I bet you're a hundred pounds soaking wet. Incest is best; put your brother to the test."

"Boy, you must be dumber than a dick! No brother either, dickhead."

"How the fuck would I know that? I just met your skinny ass. What family member have you fucked? I know these jokes are low-hanging redneck fruit."

"We do not fuck our sisters!"

"Uh-huh, but all Californians are gay based on location. Solid logic. It's okay for you to make those assumptions about me, but if I turn it around on your possibly inbred ass, you freak out."

"Speaking of faggots, did you hear about those guys who want to be sent home, so they sucked each other's dicks and made sure someone would catch them?"

"That was quite the segue. No, did that happen?"

"Oh yeah, they didn't get sent home," he replied. "They were just sent to other units apart from each other. You know, so you can't keep sucking each other's dicks. You sure you don't suck dick?"

"No, I have different hobbies than you do."

"Boy, you got a mouth on you. Someone is going to beat your ass," he said with his thick hick drawl. This guy looked right out of a coal mine.

"Not you. You look like a sentient rake," I replied.

"Did you hear about the wife who was edited in her getting railed by a neighbor in the middle of a movie? This whore sent the movie to her husband. Her husband puts it in the VCR, and everyone is enjoying the movie, then BAM! She's getting dicked down by the neighbor. It takes a second for him to realize what he's seeing, and he loses it."

"I bet. That is fucked up." I reply.

"This fucking whore sent this video instead of just sending divorce papers and ruins *Apocalypse Now* for this poor guy. He will never enjoy that film ever again."

"That's your big takeaway from the story? She ruined *Apocalypse Now?*"

"Well, I just love that movie. We have a movie tent with a small library, too. We get movies in care packages once a week. We have magazines but no recent newspapers. I mean, this guy is risking it all for his country, and she fucks Jody."

"Was that his actual name?" I asked.

"I don't fucking know, but she is fucking this guy. A videotape is like ninety dollars. Aren't divorce papers cheaper? I don't know how she even put it in the middle of the movie."

"I think you would just fast-forward to the middle, then put the tape in the camera, film the shafting. I don't think she edited between two VCRs. It isn't complicated. I get that with your West Virginian schooling, this would baffle your country ass. Do you call the TV a moving picture box?"

"Fuck you, city retard fuck," he replied.

West Virginia has cities last time I looked at a map, but these bumpkins constantly bring this up like it is an insult to live in a city.

"Eloquent. It just rolls off the tongue. I'm from the suburbs, not the city. California is not one giant city. Are you afraid that getting your picture taken will steal your soul? You're a fucking bumpkin," I said, laughing.

"I am going to fuck you up one day. Your mouth is cashing checks your body..."

"Yes?"

"Wait, your mouth is cashing..." He trailed off.

"Do you prefer *yokel* instead of *bumpkin*? I get what you're trying to say. But don't hurt yourself. I can smell your brain working. You're driving, so I don't want you to stroke out and kill us both."

We sat silently for about five miles, him shaking his head for three of those miles, and he broke the silence and asked, "Do you drink?"

"Is there alcohol in Saudi Arabia? We were told that it was against the law."

"Well, there is Saudi moonshine everywhere. We call it hooch! It's essentially gasoline and undrinkable. If you hate the enamel on your teeth, start drinking it regularly! The word is that some people have gone blind drinking it. But, of course, this is just a rumor. I'm sure it's good. I can get you some. You can't stop people from being people; they will always rebel against what they can't do. I mean, I will drink anything to escape the boredom out here, so I get why people risk it."

"I'm good. I love having the gift of sight, so I'll pass. I'm not much of a drinker. I've only had five drinks in my entire life!"

The two-lane road has sand swallowing up the road every few miles, so we have to slow down to drive over these mini-dirt hills. This adds an hour to what should have been a 45-minute drive.

"We'll see how you feel in a couple of weeks. You're going to meet Sergeant Slevens. He will tell you that you'll be the driver of this truck. That is my job. You tell him no."

"I'm not sure. Am I allowed to say no?" I reply.

"You are e-nothing. What's he going to do to you? Knock you down to e-zero? Is it a civilian? The rank does not exist! He's not going to send you home for telling him no," Wingood said.

"If that's possible, then I'll say whatever you want, but this is stick shift anyway, and I can't drive a stick, so that won't work."

"They'll teach you that some of the dumbest people on earth can drive a stick shift. But this is my truck, and I'm not giving it up, so you tell Sgt. Slevens to fuck off!"

We pulled up to our encampment, about an hour from the middle of fucking nowhere. It was just flat beige in every direction. The perimeter comprised six to eight track vehicles, all facing out and away in every direction. We were hundreds of miles from the Iraq border. Several large tents were beside makeshift showers, and the shitters were a few hundred yards from the camp. We pulled up near the shabbily put-together showers. Next to the showers was a large, empty Army-green tent with a smokestack sticking out on the top. I grabbed my duffel out of the back, slung my weapon, and put on my Kevlar helmet. Staff Sergeant Slevens exited his tent. He was thin and tall with the requisite sergeant mustache.

I swear every sergeant has the same pencil-thin to medium mustache. I could tell your rank just by looking at the facial hair.

"Murgia." He pronounced my last name as Mur-gee-ah and sounded it out. "Welcome to Bravo Company, 3rd Engineers." He shook my hand. He patted the truck with his hand and said, "You rode over in your future responsibility. So, this is going to be your truck. You'll stay in that tent until we determine which squad to put you in. Also, don't be fooled by the weather. It gets cold at night in the desert. Use your sleeping bag."

I locked my eyes with Wingood and said, "I can't drive stick, so I don't think I'd be a good fit for this truck. But Wingood here loves driving it."

"I got other plans for Wingood, private. We can teach you to drive a stick, so it's no problem. You're here to learn. The other important feature is the showers here. The shitters are way over there. You will get shit-burning duty; there is no way to avoid it. Suck it up. Stay downwind if you can." He pointed to the empty tent. "You'll have this tent all to yourself until we figure out what squad we'll put you in. It has a stove, and we store a bunch of shit in here, c-rations, MREs, and shit ton of cots. Grab a cot and set up your sleeping area. We do PT in the morning. Any questions?"

"No questions. Thanks," I replied.

He told me the rules: always wear my Kevlar helmet and never leave my weapon behind. I will be in deep shit if caught without your rifle—next, Sgt. Slevens pointed out the mess a few hundred yards away, where we got hot chow every morning and evening. We also had stacks of MRE boxes to grab whenever we needed them.

I grabbed my stuff and entered the tent, which was empty except for a pile of unused cots and a single wood stove. I set up my cot and threw my duffel underneath, and two men entered the tent. Specialist Caldwell, a six-foot-three man-child, held his requisite cup of cocoa coffee, his concoction he put together from the MREs. Coffee and Cocoa are in every MRE package. The other guy, Corporal Bard, an out-of-his-mind soldier, had a bushy mustache, unlike the sergeant's standardized one. He had a dark aura and personality. He just seemed off.

"Hey, I'm Caldwell. This is Bard."

"I'm Joe. Are you the welcoming committee?" I asked.

No one said a word for an uncomfortable thirty seconds. They stared.

"You know, Joe," Bard said, "we could fuck you right now, and there's nothing you can do about it. Typically, we call people by their last name, but I can't figure out how to say your crazy shit fuck name."

"Correction: you can *try* to fuck me right now, but, in the end, you'll crawl out of here, missing your balls," I replied. "And it is pronounced 'Mur-gee-ah,' just like it is spelled. Six letters."

We all laughed, and everyone enjoyed a tasteful rape joke.

Caldwell said he was in the 3rd Squad, and Bard was in another. They invited me to play cards with them later, maybe Uno. Caldwell was enamored with his father, who owned a court reporting business in the Texas panhandle.

"You know, my daddy..." he said in a strong Texas drawl.

"I swear if you repeat *my daddy*," I gritted out. "You are an adult man. I will only accept 'my father' or 'my dad' or 'Father.' It's just gross. I need a shower every time you say it. The term *daddy* is disgusting."

"My daddy used to say..." This was followed by some ancient proverb, probably from the sixties, and used incorrectly.

He also talked about his court reporter father as if he were the lead in a John Grisham novel. I always picture Foghorn Leghorn wearing the white Colonel Sanders suit and pontificating, "I say, I say...I got court reporting to do, boy!"

I got into my duffel, pulled out my two Ziplock bags of cassette tapes, pulled out the Walkman, put in a cassette, and hit Play. Just as I did, Lieutenant Stockton enters the tent, fresh out of the academy.

I began to jump up, but he told me to sit. He introduced himself and looked over my tape collection.

"Damn, Murgia, are you hear to fight a war or listen to music?"

"Listen to music now. When did Congress declare war? I must have missed that," I replied with a massive grin.

He looked at me. "We got a smart cookie here. Grab your weapon and your Kevlar. Please meet some of your colleagues. Here's a little warning: Check your boots every morning. Scorpions love hiding in them! We also got giant flies and many other critters that like to bite and suck your blood. Some even lay eggs in your ears or any orifice."

Our first stop introduced me to Corporal Beck, with whom I would become great friends, and his lackey, Specialist Daily. He reminded me of that parrot in Aladdin. Beck and Daily jumped out from underneath a Humvee they were working on, covered head to toe in grease. Beck was a short guy.

I used to say he was on the cusp of midgetry. Is that an English word? It is now. Specialist Daily was a skinny guy with his starter sergeant's mustache.

We moved from squad to squad, glad-handing and making introductions. Most people introduced themselves by their first name and where they were from. I got the "I hope you're not a fag" from about seventy percent of my fellow soldiers who heard I'm from California.

One of the things that I appreciated about being in the Army was the opportunity to meet fascinating people from all over the country. I encountered individuals from Texas, Louisiana, Oregon, Pennsylvania, Georgia, and West Virginia, many of whom came from low-income families. In my experience, there were no wealthy individuals among the enlisted soldiers, although some of the officers might have been exceptions. I vividly recall one friend from Louisiana who shared with me that he had never spoken to a White person before basic training. It was an eye-opening revelation for me.

I returned to my tent and saw a guy sitting on a diesel can before the entrance. His name was Gus, and he was a Native American. Beck had already pointed out the hole in the desert where Gus was living; he refused to live with anyone, especially white people, so he lived in a hole like a Hobbit. This wasn't the Shire; it was more like Mordor. Gus introduced himself with a big grin and a handshake. I asked him how things were going, and he couldn't complain. He was friendly and outgoing, not what I suspected of someone living in a hole. I wouldn't list it on Airbnb, but he seemed comfortable. He was thin with wild black hair. It was longer than Army regulation allowed, just like my hair. I Aquanetted the shit out of my hair, so it stayed down flat. I hadn't had a haircut in a couple of months. My hair grew like a weed. I had a descending hairline; it was unstoppable, like bamboo.

"Do you know who Garth Brooks is?" Gus asked.

"No, what the fuck is Garth?" I replied.

"That's his name! Do you like the Eagles?" He said, laughing.

"Yeah, they're all right. I don't want to hear Hotel California again; it is played non-stop on the radio."

"Well, he sounds a lot like the Eagles. Especially the 'Thunder Rolls.'

"Is that a song or album name? Have you ever met someone named Garth?" I ask.

"No, never met another Garth, but this guy is the shit. Yeah, Thunder Rolls is the first song on the tape; it's dark, too, unlike other country music. It's about cheating. Most country music is about losing your woman, horse, or pickup truck." He fidgets with the cassette tape, opening, reading, and closing it again.

"I'm from California. I've never heard any country music besides Willie Nelson and Patsy Cline. I think they are considered country music. I'm related to Patsy Cline by marriage. This is probably the only reason we listened to her at all. My parents didn't have great taste. The Beatles were life; we had all their 45s. I listened to them non-stop; it was all white suburban music I was exposed to. I owned the Karate Kid soundtrack, but it was not my proudest moment. I mean, the first rap I ever heard was from Debbie Harry; I'm white as fuck. Beastie Boys rule, and RUN DMC is fucking cool too. I am open to listening to new types of music, so I'm interested; I like darker lyrics. Caucassionally, yes, that's a word...it means occasionally doing things while being absurdly white. KISS, the Destroyer vinyl, was in every house in the neighborhood I grew up in."

"I've never heard that word! You are white as fuck, for sure. You're almost grey. KISS is cool, and Peter Chris is cool. Patsy Cline, huh? I know her shit. That is so awesome! I'm not related to anyone famous. I always thought I might be related to Geronimo, but I'm not—just bullshit family history. Everybody wants to be related to a famous dead Indian. I don't listen to the Beasties but know who they are; I promise you'll like this. I'm not into rap; I like country and Western music. I like heavy metal, like Metallica and Guns N' Roses."

He handed me a cassette tape. "Get the tape back to me when you're finished, and tell me what you think. I have more. I got Hank

Williams and probably more stuff you haven't heard. What do you caucassionally do?"

"GnR fucking rules! I try to listen to other types of music outside my wheelhouse: Rap and R&B. Bobby Brown, shit like that. I also listened to a ton of Jose Feliciano. Do you know him? Spanish or Mexican? I'm not sure; it's probably best not to speculate. I will give this Garth Brooks a listen. Garth? What a fucked up name. I'm burning out on the tapes I have. If I hear Belinda Carlisle again, I might end up in a fetal position. I appreciate the new music." I replied with a massive smile on my face. He seemed cool. Did I make a new friend?

It was getting dark now, and I was getting hungry. Gus and I headed to the chow line and took my food to the shoddily put-together picnic table outside my tent, next to the showers, Sleven lorded over. I devoured the entire tray of hot chow. Sgt. Slevens told me they heated the showers at 5 a.m. He controlled the hot water. He controlled the heater. He used up all the hot water for himself almost every day. It was the equivalent of a cold plunge when he had to shower. My genitals would climb back inside my body at the shock of the cold water. After dinner, I headed into the tent and got into my cot. I had no guard duty but a 5 a.m. wake-up for PT. It had been days since my last jerk-off session. I was in no mood.

I had the tent to myself, and I desecrated the area. Running in the sand, wearing boots, with a knee injury, was a bitch. It was a mechanized division. I put the Garth Brooks tape in and let the thunder roll over me. I felt my musical horizons expanding. I listened to my very first country album, and I liked it. I was surprised at how dark it was. Gus was right. Garth Brooks was a mainstay in our camp.

The other squads would play N.W.A. and Tupac. I heard classic rock from different parts of the base camp. It was like a world music festival. Corporal Peel gave me the Doors' greatest hits, a band I had never listened to.

My mom would tell me about him, but I never really heard his music. *Where have they been all my life?* I was introduced to new things every day. New drills, like putting on your gas mask. This was done daily; this was life and death, and I took it seriously. Every day, you were assigned to a different duty. Somedays, it was a shit duty, and other days, it was a less shitty duty. I would meet new people, too, and spend time talking to them about life and their lives back home. Music was traded amongst everyone, and we all had cassette players. New music would come in care packages sent to the soldiers. It wasn't a standard issue, but my Sony Walkman was my favorite piece of equipment. It had a cassette tape player and a built-in radio. Music kept me sane, and I had a Ziploc bag filled with escapism.

You want to talk music? Let me Mott your Hoople! I had cassette tapes featuring various artists, from MC Hammer to Pink Floyd, the Beatles to Beastie Boys. George Harrison's "All Things Must Pass" I even picked up the soundtracks for *Young Guns 2* and *Dick Tracy Soundtrack* from the traveling PX that visited our camp. The PX stands for Postal Exchange, a traveling Army store for troops. The traveling PX didn't visit our camp very often. I hit up Tower Records before I left home after basic training.

The Saudi government censored Madonna's cleavage. Who could dislike cleavage? The Jon Bon Jovi, *Young Guns 2* soundtrack is a banger. My parents sent books and beef jerky. My jerky levels were dangerously low. We listened to the Armed Forces radio as they did in WW2 and Vietnam; they gave us a steady stream of propaganda and music.

We had no Adrian Cronauer, no "Gooooood Morning, Saudi Arabia!" It doesn't roll off the tongue like "Gooooood Morning, Vietnam!"

I had music, books, newspapers, and magazines, sometimes months old. I read those old magazines until I could repeat them back to you. That was all we had. I read a lot as a child, but since I got a bike and video games, that tapered off quite a bit to almost zero reading unless it was assigned by school. I loved Tolkien, Steinbeck, and Hemingway. I started reading ten hours daily and fell back in love with it. There was nothing else to do.

A few days go by, and Sgt. Slevens burst through the flap of my tent. That day, I'd learn to drive a stick shift on the deuce and a half. Wingood came behind him and immediately started bitching at me for taking over his job. He told Sgt. Slevens straight up that he wouldn't teach me. They started yelling at each other, which I didn't realize you could do, still brainwashed from basic training. There were no consequences; it was a workplace disagreement. This would inform how I would carry myself throughout the rest of my time in the military.

A frustrated Sgt. Slevens grabbed another driver to show me how to drive the truck. I don't remember who exactly, but he jumped in, and I got in the passenger seat. I remember someone being in the back of the truck. We came to an open space; the entire desert was open.

I hopped onto the driver's side and spent five minutes stalling the truck. The truck was lunging and then coming to a stop. I finally clutched correctly and got to second gear. The clutch forced back and made that grinding noise every time I tried to put the clutch into gear. I heard the engine climbing. I was now in third gear. We were starting to move now. We were on a level plane, or so I thought. I was unsure how fast we were going; it felt like twenty-five mph.

We picked up speed. The road got bumpier. I hit fourth gear. I was running over desert plants and rocks, which kicked up and hit the windshield. My co-pilot never told me to ease back. I was an idiot, so I kept putting my foot down on the accelerator.

Out of nowhere, I saw what looked like a dry creek bed or an indentation in the earth, no less than a few yards away. There was no turning right or left to avoid it, and it was too late to stop. I stomped down on the clutch, put the shifter in neutral, and took my foot off the gas. I prepared for the worst and gripped the wheel tightly. My passenger also braced.

How fast are we going? We hit the creek bed hard; I felt the front axle below me bottom out as we hit the dried-out creek. I could feel the metal scraping. I'm holding the steering wheel with all my might to keep the front wheels straight. This was an act of futility. We are still moving fast. We both felt the suspension snap back upwards as we hit the other end of the raised bank; it was a three-foot berm of dirt, no more, but it acted like a ramp. The truck bed was shaking violently as we traversed the uneven terrain. There are no seatbelts. All of that stopped when we hit the ridge of the creek. We were now airborne. Now, we feel the truck's back wheels hit the barrier and launch us even further. Now, it was just silence in the truck; I felt a bit of weightlessness for a second. The front end is higher than the back end. I grip the steering wheel tightly. We are coming in for a hard landing. The front end hits hard. My chest slams into the steering wheel. My passenger's face slams into the windshield; blood explodes out of his mouth. The back wheels now touch down. The truck bounces, and all four tires never touch the Earth again. We slam down again, teetering on the left and right wheels. We are still moving forward, kicking up dirt and sand as we almost roll over. The truck lands, and we slowly roll to a stop. We sat there silently. I've only been in the country for two weeks, and this truck almost murdered me.

My driving teacher and I were shaken up and started laughing about it. We agreed not to say anything to anyone because we would both be in trouble, probably me more than him, but who was keeping score? We got out of the truck and looked for any damage. I walked back to the dried creek bed. We were airborne for at least 25 yards. We could see where the wheels first came down in the sand and kicked up a big divot. The entire story of the jump could be seen on the desert floor.

We walked back to the truck and gave it another once-over. No damage on the truck was visible, but I felt metal on metal, which felt like the bottom of the truck had been scraped in the desert. We got under the truck, with rocks lodged in different nooks and crannies. We pulled them out, but nothing else seemed broken. We both agreed he should drive us back to the unit. We return to the unit, say nothing, park the truck near Wingood's hooch, and walk away. I never heard anything about it. My driving instructor returned to his tent to put a band-aid on the cut on his head. He told a story about falling and hitting his head on a rock. I retired to the giant empty tent to lie down and count my blessings that I didn't die. I threw on the headphones and retreated to my safe space.

Sergeant Slevens had a morning routine: climb to the top of the shower where the water tank was located, light the gas heater, and keep the gas knobs hidden to ensure he got the first hot shower of the day. Unfortunately, the showers were poorly built, with nails poking out everywhere, and the water was constantly freezing when I got around using the showers. The showers were about ten feet from the opening of my tent. He did all this so he would get all the hot water. First, he didn't give a fuck about anyone else; the hot water was just for him. I heard a commotion outside and someone calling for help. I see a man hanging off the shower, the ladder lying on its side.

The man is kicking and screaming, trying to rock himself back and forth off whatever he is caught on. I realized Sergeant Slevens was hanging by his PT shorts, which were caught on a nail. He must have slipped and fallen, struggling to get free. I burst into a fit of laughter. I fell over. He looked so stupid and helpless.

Sgt. Slevens asked me to move a ladder over to help him down, but I was too busy laughing to be helpful. I just walked back into my tent, laughing. The sergeant eventually freed himself by moving his arms and kicking his legs (I'm assuming). He fell about five feet to the ground.

Slevens stormed into my tent, ripped shorts in hand, red in the face from embarrassment, and screaming. I was still laughing, unable to compose myself. He was nose-to-nose with me. His breath was so bad that I didn't know if he needed gum or toilet paper. This did not endear me to him. He is only out for himself, so I show him what it feels like, and he is unimpressed.

Sgt. Slevens would go into a local town and pick up items for the soldiers. He had a list and prices. You could buy a personal stove, candy, soda, snacks, and hygiene items. The PX truck came to our site again, but had little on it, so you relied on Sgt. Slevens.

Gus used to drive Sgt. Slevens went into town, so he knew where the Saudi store was. One night, Gus woke me at about 10 p.m. and said he would take or steal the Humvee and go into town. "Let's go to the store and pick up some stuff ourselves. We must write down the prices. He never lets me get out of the Humvee while he shops. Something is rotten in Denmark. He never lets me in the store for some reason."

We waited until Slevens was asleep. It was cold in the desert at night. I thought my genitals retracted into my body. The Humvee had no windows but plastic flaps.

"Close that pneumonia hatch! It's fucking cold," I yelled at Gus in an annoyed whisper.

"What? What hatch are you talking about?" he asked.

"The window, idiot. Seal the flap so the cold air doesn't come in."

"It's not that far a drive."

"I don't want to get hypothermia. Just seal it up!"

He sealed the window. It was not that it did much, but I felt better. We rolled a few hundred yards in neutral before starting the engine and speeding. He knew Sgt. Slevens was ripping everyone off, and he wanted to prove it.

I got some money to pick up some stuff. This was a mom-and-pop shop. This, technically, would be a tiny village, not a town—just a few adobe-type buildings. I wasn't expecting a fully stocked 7-

Eleven, but the limited selection would be generous. They had a generator that kept the lights on. I saw no refrigeration. They had off-brand everything. Camel jerky by the pound. Weird potato chip flavors. I had my pad and pen, and we wrote the actual prices of the items in the store.

Our suspicions were correct: Sgt. Slevens was up-charging twenty-five percent and pocketing the difference. This is against the military code of conduct; it is called profiteering. We had to cut the headlights and kill the engine when we returned. We parked the Humvee in the exact spot so no one would be the wiser. We didn't know what to do with the information, so we kept it secret but let our friends know they were being taken advantage of.

We were both privates; no one listened to us, the lowest of the low. You really can't trust anyone. He is supposed to buy us things at the store to make our lives easier and comfort us, but here he is, trying to profit off his troops. It was disgusting. We wrote down all his crimes and kept them, which would come in handy a few years later in my military career.

Christmas Day, 1990. It was my first time away from home during the holidays. I wasn't religious, so I enjoyed the break from school and the presents. Christmas music played, but there was a massive sandstorm that evening—another first. You could see it coming; it was a wall of sand, and it must have been eight stories high.

I appreciated them making us forget we were in the desert at Christmastime. The more you tried to make it feel like home, the worse I felt. The howling wind, rocks, and sand hit the tent, but it was relaxing, like a soothing white noise. It sounded like rain hitting canvas, but it was tiny rocks.

Gus was not in his hole during the storm; he was hanging out with me in the tent. A few other soldiers were there. We played cards and stayed warm. We watched the lieutenant go from tent to tent with his shaving cream Santa beard and hat. He had a sack over his shoulder and would give everyone gifts, a can of soda, and a candy bar. His white shaving cream beard was filled with sand and rocks when he got to my tent. It looked like he had a beard of bees on his face. It also began to melt and drip off his face from the stove's heat.

"Ho, ho, ho!" he exclaimed.

We broke out into laughter when we saw him. We had to hide the Saudi moonshine we were sipping on to make the day tolerable. He tried spraying a new layer of white cream over his bee beard. He handed out his gifts and talked to us about his family. It was excellent, and I appreciated his effort to make that night less weird. All the days, we blended into one another, and on the night, we ended with our group singing "Friends in Low Places" at the top of our lungs. It was like a fucking Nora Ephron movie.

I also learned from the lieutenant that I was heading to the 3rd squad in a few days. He told me the 3rd squad was the best fit since he saw Caldwell and I got along. Caldwell was the track driver, so our track vehicle was designated one three (1-3). Our squad leader was Staff Sgt. Jess Steward. The youngest E-6 excelled at everything he did, and I was repeatedly told what a decent man he was. This was the squad I wanted to get assigned to since Caldwell also boasted about Sgt. Steward. I was thrilled to be part of a squad with a leader everyone in our unit respected. I loved being in that giant tent by myself. I had no guard duty. Being part of a squad means guard duty, and it's a drag. This meant getting up at some point in the middle of the night for an hour and making sure no Iraqis attacked us in the middle of the night, even though they were at least five hundred miles away. We also had to get up at sunrise, sit in our foxholes, and wait for the lieutenant to come around and check on us. The morning desert ground was cold, and I wore airborne boots as if they were metal plates. Leaving your boots on the ground was like putting them in a freezer overnight, plus the scorpions, which were always looking for them. I was getting used to living independently in my tent; I could masturbate freely, and I got to sleep through the entire night. It was a gift, but now I had to be part of a team. The team that would help me make it through whatever came next.

CHAPTER 5

I packed my gear and headed down to the 3rd squad. I finally felt like I was a part of the unit. I'd made friends with a few guys, but being part of a squad was different. Caldwell showed me around a bit, and we headed over to Sgt. Steward was sitting on his cot, reading a book. His cot was adorned with Simpsons sheets and pillowcases. Tall, thin, blondish brown hair, sergeant's mustache. He had a thick southern drawl. He pronounced the word *creek* as *crick*. Caldwell never left our side while Sgt. Steward took over the rest of the 3rd squad tour. Sgt. Steward looked me over. I looked fourteen.

"Jesus, how old are you?" Sgt. Steward asked.

"Nineteen."

"Well, our squad is tight. I've heard good things about you from Caldwell. You also seem to enjoy annoying him, bringing a smile to my face. Now, this track is 1-3. We each make our hooch to sleep in or around the track vehicle, grab a cot and pick a good spot under that asshole camo net. Lay your sleeping bag on the cot and store your duffel underneath. It rarely rains, so you don't need a poncho covering you. This is where Bunk and Caldwell sleep. Over there in his little tent away from everyone is Corporal Peel."

That was how he pronounced Pill. He even elongated it to *Peeeel*.

"Corporal Peel is seemingly allergic to showers, so we don't mind him being a bit away from us. One time Hell took off his socks, and they stood up on their own like they still had feet in them; I shit you not. He also ordered a video cassette recorder and a camera and shipped them out into the desert. Can you believe that shit? The camera arrived here without any issues. I can barely get mail at home! Fucking Fed Ex'd it! Peel is going to make a shitty documentary or something. I'm not sure what he's going to do with it. Useless, if you ask me,"

Steward continued, "This is history, but he is out of his mind. This track vehicle, or APC, stands for Armored Personnel Carrier, the M113. Coincidentally, we are Bravo 1-3. We call it the track, so don't overthink it. I want to draw your attention to the camo net covering said track. You will learn to hate that mother fucker. We must take down and pack that camo net every time we move. It never rolls up right. You will hate that camo net more than the Iraqis by the time this is over. Everyone, say "Fuck you, camo net.""

We all did, in unison, like *Stepford wives*, "Fuck you, camo net." We then moved on. He was right, fuck that cargo net. I have never encountered a more infuriating inanimate object. It would never fold up correctly, so you hold it together with straps. Finally, you would have to fold and unfold it before figuring out the magical way to strap it together and load it on the track. It is like changing a fitted sheet on your bed while being high, borderline maddening.

We approach a small army-issued tent just a few yards from the track vehicle, "Peel said. Peel, you in there? Tell Joe how you, Astro, produce yourself in Germany every night or whatever you do. Sounds wild to me. Hello? Is this bitch in here? I don't know. I don't know where he goes. Do you watch *The Simpsons*? As you can see, my bunk has Simpson sheets and pillowcases. Unfortunately, I don't have any extras, so please don't ask. Do you know the show?"

"No, I haven't seen it yet. I was in basic training; we had no access to television or soda. I think Peel said astral projection, not..." I replied.

"I don't know what he means, it is some crazy horesshit, he doesn't do. It's called dreams, asshole! You go nowhere!" Sgt. Steward turns his attention back to me.

"You just came from the world. How do you not know *The Simpsons?*"

"I just got out of Airborne in September. I then went to Airborne school, but haven't watched television in almost a year. I've seen the Tracy Ullman show." I replied, thinking he knew the origins of the Simpsons.

"I don't know what that is; I never saw it. Over there, we got Bunk. He rarely gets off his cot. We don't ask much of him."

"That's fucked up," Bunk responded from under a blanket.

The ramp on the back of the sand-colored track was down. Two makeshift boxes lined each side of the track. There were no seatbelts or safety features.

"Now, in the back of the track, we have two wooden boxes." He said.

"You sit on these uncomfortable pieces of shit—sorry, no padding for your delicate ass. On this side are tools, shovels, and shit. You will open this box a lot. We got picket pounders, crimping tools, and all sorts of bullshit to do your job. This other box, though, is nothing but explosives. We have C4, and this dynamite is leaking something fierce.'"

"This looks unsafe," I said.

"Oh, it is. Leaky dynamite is nothing to fuck around with, and we got a lot of it. We have this box. Be fucking careful. If you want to grab some of this leaking nitroglycerin and fling it at Caldwell, use your knife edge, but do so sparingly. I'm not going to bust you. He probably deserves it. You'll figure it out. We have C-4 explosives next to the dynamite. We received a box containing blasting caps, detonation cord, and timing cord. You will never handle these—just me. I will give them to you as needed when the shit goes down. We got a det cord and a timer cord. You remember what to do, right? If not, I'll provide some refresher courses and demonstrate how to conduct a test burn. I mean, it burns differently depending on the temperature. We'll go over all that stuff. Another thing: close that box, Murgia. Don't mess with this side of the box. I'm the TC. We only have a couple of pairs of night vision, Caldwell and I. Caldwell drives me crazy. I can't explain why. He annoys me most days. Any questions so far?"

"Do we have bullets yet? This appears to be a crucial aspect of the war. Are they in the track?"

"Nope, no bullets, not yet. They don't want to deal with armed teenagers, you know, no offense. Do you even have hair on your balls yet? Look how smooth you are, not a line or hair on your face. This war might age you up quickly if we even go."

He continued, "I don't think they will give up, so be prepared to go. I don't know when, but it will be soon. We are going to shit or get off the pot at this point. A lot of money is being spent, and you can believe the U.S. government is getting bang for its buck. Look at all this money just sitting around. All the paychecks and the equipment. Now, over here is Johnson. You'll be in the foxhole with him tomorrow morning. Now you gotta do things by the book, so when Stockton walks up to your position tomorrow, follow Johnson's lead. I don't know where Heel is, so I gotta figure that shit out now. Now, my right-hand man, this is Sergeant Martinez over there. Have you met him?"

"No, not yet," I replied.

"Great guy, great non-commissioned officer. He got kicked off *Romper Room* for telling a joke. I don't know the joke, but I'm sure you'll hear it. He's from the East Coast. Philly. He fucking peaked at five. Great guy and a decent man. He'll take care of you. We watch each other's backs when shit hits the fan. I heard you're from California. What part?"

"San Diego."

"California, huh? Do you surf?" he asked with a big smile.

I sheepishly replied, "No."

"No? How the fuck do you live in San Diego and not surf?"

"Just because I live in California doesn't mean I surf," I replied. "But yes, I get it." I'm not the biggest beach guy. I'm not too fond of the sand in your crack. The fucking water is cold as shit in Cali, it is not like the Atlantic Ocean, and there are rocks and coral everywhere, too. The East Coast has warm waters; you can walk out several miles, and it's still only up to your waist. It isn't fun to me. I sometimes enjoy lying on the beach, but that is about it."

"A non-surfing Californian? You're a fucking mystery. Do you know the chow schedule? I'm sure you do. We train in this squad. You will teach things, too. You'll figure it out. We practice putting on the gas mask daily. I will sometimes surprise you guys. But, of course, you do not want to be gassed.

"We do guard duty for an hour every night. I expect you to take it seriously, even though we're five hundred miles away from Iraq. We'll ease you into a couple of easy shifts. You are the lowest rank on the totem pole, but we make our picks fairly. You may get the shit shift, but I try to be fair. Good work waking Bunk up. He might be narcoleptic, but I'm not sure. I've never seen someone sleep so deeply. It may take you twenty minutes to wake his ass up. We used a mirror near his nose to check if he was still breathing. Now, go back to classes and think about what you can teach. It can be anything. Do you have a particular MRE combination? Teach us. Last but not least, the MICLIC (pronounced MICK LICK). It is spelled M-I-C—"

I interrupted, "K-E-Y?"

He stopped. "Quit fucking around, Murgia, and listen up. M-I-C-L-I-C. You got the box here filled with a spiral of C4 explosives. I believe each one is half a pound of C4 explosive. They look like little C4 sausages. They must be laid in the box in a very particular way.

You don't want them getting tangled up when we launch that missile. You want them lying down in as straight a line as possible.

We got the arm that raises pneumatically. Do you know what that means?"

"Of course. Air."

He nodded. "Correct, air. We have the missile on the arm, and it's attached to this coil of explosives. We launch the rocket, and it pulls the C4 out of the box, and it lands on the other end of a minefield. Now, all that C4 is lying across the minefield, too. So, we set that off from inside the armored vehicle, and it blows a lane wide enough in that minefield so vehicles can safely cross. It's simple, but we're the squad that takes on that responsibility if they call on us to do it. So don't worry, we'll teach you everything you need to know."

"I'm not worried in the least," I replied.

"Good. I also recommend that you read. You got lots of time, so educate yourself, and don't spend all your time fucking around with leaky dynamite. Well, that's about it. The mail call is in a bit. I'll take a nap, so don't bother me until dinner. If you see those sheets over my head, go find Martinez and bother him."

We circled the entire squad and ended up back at his hooch. He looked me up and down again before retiring to his bunk for the afternoon. "Jesus, you're just a kid. Stick by me, and we'll get through this together. Murgia, I feel like I owe it to your parents to get you out of this. Just do what I say, and you'll be fine. When you do get home, you really should learn to surf."

"Okay, you're hung up on that."

"Now go set your shit up," he said as he threw his Simpson sheets over his head.

I headed to the side of the track, stretched my poncho out, and set up my cot underneath. I lay in my sleeping bag. Sgt. Martinez walked up to me as I was smoothing everything out. The rest of the squad was not paying attention to me and was just in their own worlds.

"Hey, I'm Sgt. Martinez. You can call me Sgt. Martinez. How are you, Joe? I see Steward gave you the rundown. Everyone is cool in the squad. Caldwell is annoying, but I'm sure you figured that out. I heard him say *Romper Room*. Did he tell you my joke?"

"No, he couldn't remember it. It sounds like the greatest joke of all time." I replied sarcastically.

"He couldn't remember it? I'm glad; I'm probably the only one who can do it justice! It's a great joke." Martinez said with a wry smile.

"He didn't tell me. How great could it be? Is it a kids' joke?" I asked.

"What do a stove and a woman have in common?"

"Is this the joke?" I wondered.

"Yeah, dummy, do you know?"

"No, I have no idea," I replied.

"Both have to be hot before you put the meat in." He started cackling to himself. "The whole station went off the air after I said it. Some teenagers encouraged me to do it. I was removed from the studio. It was great. I'm sure someone has this footage. It's a dead day; we've completed our training. Chill out for a while and get some shut-eye before chow. I'm putting together the shifts for tonight. We'll walk over and get chow later."

But as soon as I saw Sgt. Steward hit his bunk and pulled the sleeping bag over his head; a sense of calm washed over me. He exuded confidence and was undoubtedly the alpha of the group. His presence commanded respect from everyone around him, and I was no exception. His cocksure attitude made me feel like I was in good hands and safe.

His concern for the squad and me was palpable, and it was evident that he would do everything in his power to ensure that we all returned home safely. Despite the dangers that lurked ahead, I felt an unwavering sense of trust in Sgt. Steward.

I was grateful to have him as my squad leader, and I knew that with him leading the way, we would have the best possible chance of completing our mission successfully—by that, I mean not dying. He is the best non-commissioned officer I have ever met.

We had guard duty every night. The dream shift was the first shift from 8 p.m. to 9 p.m. After that, you get the whole night to sleep. I rarely got this shift; the schedule needed to be organized. I seldom left my sleeping bag—threw the headphones on, and escaped this lonely reality. Now, I was less than ten feet away from my fellow soldiers. Masturbation is no easy task in a hooch with dudes everywhere. I needed to be milked.

Nevertheless, I'd been having at myself for the last four years, thanks in no small part to Justine Bateman on *Family Ties* and Alyssa Milano on *Who's the Boss*. And who can forget Elvira, Mistress of the Dark? I saw Elvira's movie with my mom at fourteen and had a sexual awakening, and we were only an armrest apart from each other—one of the most uncomfortable situations of my life.

At nineteen, I was an ejaculate fountain—the worst type of fountain. I was in my prime, horny twenty-four-seven, even while sleeping. I glazed everything I could, like a Cinnabon. I wasn't going to let a little war get in the way of me pollocking the inside of my sleeping bag as much as humanly possible. I had to find the correct times, usually in the middle of the night, during and after guard duty. I had no shame.

I needed that post-nut clarity. I couldn't even deal with humanity unless I got that poison out of my system. That sleeping bag should've been thrown in a volcano after my time inside it. You had to be quiet. I had no lubricant or lotion to aid me; I just had a bit of spit and a dream. I made all critical decisions in that post-nut bliss.

There was also a rumor that they put saltpeter and MREs in the food to prevent erections. If that was true, they did not put enough in. I powered right through it.

Our unit built these temporary wooden toilets. We could load them on a trailer and move them quickly. Was Sgt. Goebbels in our company? If so, that fucker oversaw putting together that splinter contraption together. More nails than necessary, most sticking out at an angle. Jagged edges, nothing sanded down.

They had nailed down toilet seats over a sloppily cut hole, with all the holes being different sizes and without uniformity. All you heard in basic training was that uniformity was critical to survival. *What the fuck does that have to do with anything?* Underneath the hastily nailed-down toilet seats were metal containers to catch all human waste.

It was a defecating dexterity test in the middle of the night—no toilet separators. I prefer not to talk to anyone while I'm doing my business. You were inches away from someone else dropping wolf bait, almost thigh to thigh—a waking nightmare.

I don't want someone speaking to me while pooping, much less be shoulder to shoulder with my fellow poopers. If it is just number one, the desert is your toilet. I marked my territory everywhere I went.

Number two, you had to head to those shitters. It was sometimes easier to walk behind a dune, dig a cat hole, and do one's business. The less time in that tetanus box, the better.

I would rather be part of a human centipede than burn human waste again. One of the worst experiences in life is having a job that involves burning human waste. You needed gloves, but it still didn't feel like enough protection. Someone would inevitably line it up incorrectly underneath the hole so that shit would hit the sides of the can, instantly meld, and become one with the can. As a result, the entire outside had a thin, caked layer of feces. No one could avoid this duty. It had to be disposed of. The only way was to burn it.

You had to drag a diesel can out there. You did not want to be downwind. It didn't matter that smoke followed me around. It was like a sentient cloud that wanted to see how much it could make me vomit in thirty minutes. I became a shit sommelier. I was picking up notes of chipped beef, *Tobasco* sauce, and, of course, my vomit.

They gave us no protection. I was told to wrap a T-shirt around my mouth and nose. The smoke was almost sentient, following me wherever I went. Did I molest shit smoke in a past life? Is that even possible? That did nothing; I could feel it in my nose and lungs for several days. Some guys wore their gas masks. They were more intelligent than I was. Never look in the can.

You can't unsee it. It was a cauldron of horror, brown and black logs, solids and liquids swirling in harmony, green rivers of viscous ooze. A gelatinous armor would form on top, which you had to break up with the "shit stick" so the diesel would penetrate deeper and burn up everything.

The nauseating aroma would slap you across the face when you first broke through that shell with the stick. Based on what I saw and smelled, most people wouldn't live to see the war. Was it the MREs? Was it eating Union rations from the Civil War? Most likely. These meals were old. It seemed unnatural that they would be good to eat. Once you stirred the shit stew to the proper burning viscosity, dump the diesel in and give it one more mix. The diesel-guano bouquet filled your senses. It can cause dizziness under certain conditions.

Throw a match in. Then, the fun began. The air became thick. It would be best if you worked on breathing. It didn't matter how far back you stood. It would find you. It stuck with you, literally. The smell got in your uniform. You'd want to throw your uniform in the inferno; the scent was *not* coming out. We did not have washing machines. I had to slap my laundry against a rock like a troglodyte. No matter how often you washed that uniform, the smell always lingered. Eventually, you would have to burn the uniform and buy something new.

After my poo-burning experience (I performed this duty no less than six times), I spent my first night with my new squad, undergoing my indoctrination into being part of a team. I had the worst guard shift that night. I could never go back to sleep. I woke up that first morning in the third squad.

"Did anyone see that coyote?" I asked.

"There are no coyotes in Saudi Arabia, you moron," Bunk said.

"Well, I think some animal shit in my mouth. I have to brush my teeth."

Every day was the same, with training and other errands mixed in to break up the monotony. We showered once a week at the most. I had to figure out which way the wind was blowing when brushing my teeth so it didn't blow sand on my brush—just annoying little things about living outdoors. I mostly read my books, old newspapers, and even older magazines.

There was pornography around, but not as much as you would think—a *Playboy*, a *Penthouse*. But, of course, the type of pornography you read says a lot about you. *Hustler* magazine was around, but that was like looking at an anatomy textbook—not my thing.

I suggested a class on napping, otherwise known as "eyelid maintenance." Sgt. Steward would constantly shoot down nap time. One lesson I remember giving was making a chocolate treat using the cocoa in the MREs.

We had to engage in culinary experiences, such as when they fed us camel burgers one day. It's a gamey, loose-meat sandwich. It wasn't good. It did *not* taste like chicken. The camel meat fell apart; they didn't use a binder, and it was grey, like Arby's. Camels are straight-up assholes. If pedophilia had a flavor, it would be camel meat—awful taste. Tabasco did nothing to hide that gamey flavor.

Every morning, we sat in our foxholes with another soldier. For me, it was either Bunk or Johnson. We talked about movies and music. My feet froze because of the metal plate in my jump boots; the cold desert ground cooled the plate down. I would get up and stomp my feet to restore blood flow.

Lieutenant Stockton walked to each perimeter position in the same pattern every morning. We watched him while we talked. We were supposed to challenge him with our security word as he got closer.

This changed daily, and he would have to respond appropriately before we let him approach. Once we challenged him, he would comically flail like he didn't know it was coming, then come to a stop and look down at us in our hole.

After the foxhole, we would head to the mess for a hot breakfast, and on some days, we would have MREs. The MREs might not have been the most gourmet of meals, but I had a soft spot for the dried beef patty. There is no Michelin star for ready-made meals. I would carefully cut into the canned bread, which always had an unnatural brown hue that made me wonder what kind of sorcery went into its creation. Yet, despite its unusual appearance, the bread was surprisingly soft to the touch, and it could be left out for weeks without going stale. It was almost like black magic bread, defying the laws of nature.

As I added water to the beef patty to reconstitute it, I couldn't help but get creative with my meal. I turned it into a simple hamburger, adding Tabasco to my rations. The single-serving bottles were cute but incredibly useful in turning a bland meal into something almost like a luxury.

Caldwell and I spent most of the days giving each other a hard time and busting each other's balls—a fun male pastime. Caldwell and I were assigned duty in the mess hall run by Sgt. Ramirez, a connoisseur of the word "faggot." I clutch my pearls when I hear it now. Ramirez was from the Caribbean and was a master shit-talker. His accent made every nasty thing that came from his mouth sound like poetry. It was called KP duty. This was a standard kitchen duty everyone had to do. It was like when Abbott and Costello went into the Army, peeling potatoes.

We walked up to the kitchen, talking. I saw a table with a massive pile of potatoes at least four feet high.

Ramirez was on the other side of the table. "You two, get peeling!"

When I got close to the table, I pretended to trip. I spent my childhood doing stupid pratfalls for fun.

Before wanting to be a soldier, my second choice was to be a stuntman, followed by a private eye—just like David Addison, played by Bruce Willis on the Television show Moonlighting, my favorite show at the time. Caldwell laughed. On my way down, I hit the palm of my hand on the table, making a loud bang. I got up, holding my head, feigning an imaginary injury.

Ramirez came running over. "Oh, my God. Are you okay?"

"He's fine," Caldwell said.

I put on a show for Ramirez. I couldn't believe he took it so seriously; he usually didn't care about anything.

"Let me get you a cold pack."

He started to run off, and Caldwell said, "He's faking it. He hit his hand, not his head."

"Shut the fuck up, faggot. I saw everything!"

"You saw nothing. He's faking."

"Listen here, faggot, you grab some potatoes and start peeling. Put the peeled potatoes in this container with water when you're done."

"He's fine," Caldwell insisted.

When Ramirez looked at Caldwell, I smiled at my friend while I knowingly fuck him over.

"See? He's faking!" Caldwell shouted.

Ramirez turned back, and I turned on my Oscar-caliber performance.

"You sit in the shade and watch Caldwell peel those potatoes. Then, let me know when you're feeling better. Then, you keep peeling them taters, faggot!"

That was genuinely how he spoke daily. I may be underusing the word. The death stare I received while sitting in the shade with my cold compress on my fake head injury... I would taunt Caldwell and tell him to peel faster.

"I'm going to murder you," Caldwell said.

"You wouldn't murder an injured person, would you?" I reply with a smile.

"*He's not injured!*" Caldwell screamed.

"Are you a medical doctor?" I asked.

He looked through me at this point.

"Peel slower. All your movements are making my head injury worse."

We had so many days of doing absolutely nothing. I would just read all day. However, we would then have other days when Sgt. Steward yelled, "Gas!"

We'd all jump up, grab our gas mask bag, hold our breath, rip the mask out of it, and slide it over our heads as quickly as possible, then tighten the straps.

Sgt. Steward would count the seconds audibly: "One, two...You're dead if I get to ten and you still don't have your mask on."

This happened frequently, sometimes in the middle of the night. You never had your gas mask out of arm's length; it was life or death. We became so proficient at putting that mask on that we could secure it in five seconds. That would possibly save our lives one day. Our team was enrolled in combat medic school for a few days.

Someone in the battalion arranged for this training so we could learn basic first aid. Clapton, our medic, was the teacher and someone I did not recognize. I believe he was brought in from outside our unit to assist. We learned how to give injections. Inserting IVs was fun. I teamed up with Caldwell, whom I called Nurse Mengele. He couldn't find a vein, using my arm as a pin cushion. They couldn't even feel me putting the needle in; I was great at it.

Next, we learned how to dress bullet wounds and deal with sucking chest wounds. This is if a bullet collapses a lung. You cover it with plastic and tape it to help form a seal around the lung, allowing the victim to get air. Bullet wounds are nasty. The M16 bullet tumbles in the air, and if it hits you on the shoulder, it can exit out of your lower back. They would show us photos of bullet wounds, which were unpleasant, but seeing gore was not too shocking since I read *Fangoria* regularly. The actual injuries looked fake, so it never bothered me.

Boredom would set in quickly around the camp. You can only read so many magazines or books. We trained on all the equipment we would use in the field, including the old radios with a hand-cranked generator. The radios were wired, and we would run the wires from squad to squad. We also had gas alarms installed on the perimeter edge.

You would have to wrap the wires around these old-timey connectors on the radio. Some people would spin the crank and run electricity through the cables while you were trying to connect them to the radio and shock the shit out of you.

Pranks were a significant way to blow off steam. Bored soldiers would also resort to horrible things to fill the time. They would capture scorpions and giant, hairy spiders and have them cage match and fight each other. Bets would be placed. It was a bug Octagon. I saw a couple of other soldiers capture these giant scorpions —

monster bugs —seven to eight inches in length. These bugs are straight out of a nightmare.

I'm not a fan of bugs, but they are living creatures and don't deserve to die just because soldiers are perverse and bored. These soldiers would wrap the wires from the radio around each end of the scorpion. They would then turn the power crank, and these scorpions would explode, leaving scorpion guts everywhere. It is not something I would ever do, but I witnessed it more than a few times.

We often changed camps, which was a giant pain—packing up and resetting everything elsewhere. The amount of garbage we left behind was unfathomable. We would hear stories of Bedouins—nomadic people, outdoorsy folk—living in the desert and traveling for centuries. We would see them going around our perimeters.

They only got close to us one day. I was sitting on a diesel can and shooting the shit with Sgt. Martinez. In the distance, I saw a Bedouin, just one, on a donkey. He was a little dot in this giant desert. It looked like *Lawrence of Arabia*, the heat coming off the desert floor. He drew closer and closer, coming into sharper focus. His donkey was packed with survival items.

"Get a load of this guy," Martinez said. "He thinks he's John Wayne."

"I'm not sure he knows who John Wayne is or any concept of John Wayne."

"This guy is infringing on his John Wayne copyright," Martinez joked.

We laughed, and the Bedouin approached us and hopped off his donkey. He was wearing a black thawb, which is their traditional garb. He wore a red checkered hat, reminiscent of a tablecloth from an Italian restaurant. It was a man's gown. I wanted to wear one to see how it kept them calm, but he only sold rugs. He waved. He was no threat. He grabbed some rugs off the back of the donkey.

"You like? You like?" the Bedouin asked.

I don't know what I bartered, but I got the rug. Our first sergeant, Winston, ran out to get the Bedouin out of our perimeter. He started yelling at both of us. Winston looked like John Wayne, standing at six feet three or four, a Vietnam veteran. He had seen some shit.

The Bedouin was unmoved. He slowly packed up his donkey and strode out of our perimeter. This was their desert, not ours, and he would come and go as he pleased. I believe my mother still has this rug somewhere in our home. It was a prayer rug, so it will never be used for what it was designed for.

Our unit had been tasked with an enormous responsibility: constructing a range and putting on a show for the bigwigs. It's a typical Army dog and pony show. This included the King of Saudi Arabia, King Fahd, and General Schwarzkopf, the man in charge. Ole Stormin' Norman. We spent days building the range. We created a fake minefield and even posted danger signs. Another part of my unit built the scaffolding for the seating area where everyone would sit.

Sgt. Steward was worked up about the show. "Boys, I have no idea why we're wasting our time doing this, but here we are. Now we're nearing the end of the show. They are spending a lot of money on this war and want to see it put to good use. We will not let them down. We focus on our one job: setting off that MICLIC.

We have practiced this repeatedly. We know our jobs, and we are good at our jobs, don't forget it! There is no need to be nervous. This should be muscle memory now."

He continued, "I'm proud of all of you, men. I mean, General Schwarzkopf will be watching. The world's media will also be present. You don't want to embarrass yourself in front of CNN, NBC, ABC, or CBS, do you? You know your job. Please do it."

He continued, "We are on a clock, too. We got a five-hundred-pound job being dropped after we cleared the lane in the fake minefield. We've got to set it off and go! Now, they will not drop the bomb on us if something goes wrong; they will circle, but that will waste everybody's time, so let's get this right on the first go. There are no real stakes here, so don't fuck up. Got it? Let's do this."

The range was filled with activity. Humvees darted around, soldiers shot blanks while running and playing war, and charges were being set off. We were scheduled to appear at the end of the show. The big finale was dropping a five-hundred-pound bomb. A10 aircraft strafed the desert with their Vulcan cannons, hitting old equipment set up in the desert as targets. Apache helicopters fired at targets we built.

Watching all your hard work destroyed in seconds by missiles and gunfire is always fun.

We drove to our pre-designated spot and waited. I peeked out of the hatch to see where the dignitaries were. They were miles away from the minefield and wouldn't even see all the details we had put into it.

Over the radio, I heard, "Go! Go!"

The track lurched forward, and we were off to our spot. We were being jerked around a lot more than usual. Caldwell was driving like a maniac. We stopped.

"Raise the arm with the missile," Sgt. Steward ordered.

The arm rose without a hitch.

"Get ready to launch." A beat went by. "Launch!"

We heard the missile take off. The track shook a bit. Smoke from the launch came through the open TC hatch. Sgt. Steward lowered his seat with Caldwell; they were inside the track when the missile went off. You could hear the coil of explosives being pulled out of the box.

Thirty seconds went by, and it was quiet outside. Sgt. Steward lifted himself out of the hatch with binoculars to verify the explosive lying across the minefield. Then, they got the clacker ready. A small box created the electric charge that ran down the line and triggered the blasting cap.

"Joe gets to set it off! Hand him the clacker." He shouted.

I hit the clacker, and nothing happened. I tried again. Maybe I didn't clack hard enough. We tried one more time, but it didn't go.

"Joe, just run as fast as you can!" Sgt. Steward yelled. "Open up a hole on the last explosive with a crimping tool, stick the blasting cap in, pull the fuse, and get back here as fast as possible."

This timer and det cord were set up prior as a plan B in case we had to set it off manually. I jumped out and headed toward the track, where I found the explosives. I heard a plane overhead. I dug a deep hole in the explosive and inserted the blasting cap. I sealed everything up tight. I ensured the girth hitch on the det cord was tight to the timer cord and pulled the fuse. I ran back to the back of the track, jumped in, and yelled, "Good to go!"

Our track vehicle lunged forward and headed in the opposite direction of the MICLIC. It only had a two-minute timer. We were a mile or so out when it went off. We all opened the top hatch and watched the show. First, the dirt showered the desert floor. There were no flames. You could only hear the bomber overhead; it was too bright to see.

The five-hundred-pound bomb was dropped and detonated.

You saw the blast, and a second later, you felt a concussion and a booming sound. It felt like it shook the entire world.

CHAPTER 6

O n January 17, 1991, we met in one of the larger tents early in the evening. Most of the company was there. There were murmurs all day that something was happening. Sgt. Steward had been pulled into meetings all day. Finally, our captain quieted us all down.

"Everyone needs to sleep in their MOPP suits tonight. Every night from here on out until we say otherwise. Understood?" he asked.

I turned to the guy beside me and said, "That ain't good."

"Yea, no shit," he replied.

In all fairness, it was a dumb reaction to the news. MOPP stands for Mission-Oriented Protective Posture. The suit has a charcoal lining to prevent gas from seeping into it. You typically wear it over a uniform.

I got the shit guard slot again, the 3 a.m. to 4 a.m. spot. We were still a couple of hundred miles away from the Iraq border. I was on alert, but not so vigilant that I wouldn't listen to music; this was standard operating procedure while I was on guard. It was a perfect, cloudless night. The desert at night, far from city lights, is truly spectacular. Our Milky Way is visible, with the sky filled with clusters of stars and planets that are viewable with the naked eye, at least with the nineteen-year-old versions of my eyes. I used to read by starlight; that is how bright the Milky Way is in the middle of nowhere—clusters of whites, pinks, blues, and purples. You sit underneath its majesty and feel your insignificance in this universe. You realize how small you are. It felt like you could see other galaxies. I imagine the same feeling you get seeing the Northern Lights in person.

A parade of planes in the sky, their formations silhouetted against the bright starlight. It was something to behold, and I'm glad I got to see it before I shuffled off this mortal coil. I can close my eyes and see it and wonder if anyone else had the same experience.

As I lay atop the camouflaged military vehicle, I watched a private show by the US Air Force and the Universe during guard duty. Above me, the sky was transformed into a dazzling spectacle as hundreds of planes and jets soared through the air, engines roaring and propelling them forward at breakneck speeds. Despite their differences in altitude and size, they moved in perfect unison, creating an otherworldly display that seemed to transcend time and space.

As I gazed up at the planes, their sleek frames contrasting with the velvety backdrop of the Milky Way. The B-52 bombers caught my eye, their massive bodies moving slowly and steadily like behemoths against the starry sky. I couldn't help but marvel at their sheer size and power. I climbed onto the vehicle's top, my heart pounding excitedly as I positioned myself to get a better view of the spectacle above. The camo netting covered most of the way, but a large opening allowed me to take in the entire scene. As I settled in, I rummaged through my bag of tapes and found the perfect accompaniment to the breathtaking sight before me: Pink Floyd's *The Dark Side of the Moon*.

The music filled my ears, drowning out the sound of the planes and transforming the experience into something magical and surreal. It is the best thing I've ever seen. The world fell away. I escaped. Just like Peel, except I didn't beam myself to some shitty club in Germany. I don't know where I went—my magical, mystical tour.

All I needed to escape was my Walkman, Floyd, and an air campaign. I lay there, completely entranced, as the haunting melody of "Breathe" washed over me. The world did not exist. I wasn't in an endless sandbox. I could be anywhere in my mind. The war had just begun, and I knew that my unit would soon be on the move. Shit, my life just started. I was nostalgic for a life that had yet to begin. I hoped the air war would be enough to bring the conflict to a swift end, but I knew there were no guarantees. The sheer number of planes, bombers, and jets was overwhelming, each at different altitudes, creating a living screensaver. I strained to count and identify all of them, and that's when I saw it—a stealth bomber gliding effortlessly across the sky.

It's a giant death triangle.

My guard shift flew by, and I remained frozen, entirely transfixed by the display above me. I watched the planes move precisely, each playing a critical role in the unfolding air war. I was free, then I looked at my Casio watch. And then, suddenly, on the horizon, flashes of light lit up the sky, revealing that we were bombing the enemy with incredible force. I needed to try now to wake up my narcoleptic guard relief.

"Hey, Bunk, look up at the sky," I called out to my sleeping relief. "The air war started, and shit is about to get real."

Bunk didn't seem impressed. He was barely conscious, his eyes struggling to stay open. The coalition forces used an impressive array of aircraft in the air war, each with unique capabilities and uses. Fighters like the F-15 Eagle and the F-16 Fighting Falcon were essential for air-to-air combat, providing air cover for coalition forces on the ground.

Meanwhile, bombers like the B-52 Stratofortress and the F-117 Nighthawk were used for strategic bombing, destroying Iraq's military infrastructure and targeting key command and control centers.

I knew none of this at the time. I woke up and turned on the radio, and I was surprised to hear the announcer say, "The war in Iraq is brought to you by McDonald's!"

It was a bizarre statement, and it struck me as funny, given the gravity of the situation. But, of course, it wouldn't be an American war without some capitalism and consumerism thrown in for good measure.

But as the days passed and plans were finalized for the attack, my squad and I listened intently to the radio for any news. We all hoped the war would end quickly and we could return home. However, the sky remained filled with planes, a constant reminder of the intense battle being waged overhead. Despite the overwhelming force of the coalition forces, a sense of unease and uncertainty lingered in the air. On January 17, 1990, Desert Shield was the operation's name before the air war started, and it has now entered a new phase called Desert

Storm. The latest phase is the beginning of combat operations. My unit would soon begin to head toward Iraq. The air war will quickly transition into a ground war. It was thrilling and scary, but no live ammunition was used. I would feel safer with bullets.

Chapter 7

W e headed north. That was our final push before entering Iraq. This must have been February 21 or 22, 1991. We set up camp. Things were serious now. There was a change in the air. Things were about to get real, and everyone could feel it. We heard that they were going to be giving us live ammunition. They had kept it from us until then.

We did not set up a hooch near the track; we were instructed to dig foxholes around the way—a perimeter inside a perimeter. Sgt. Steward made sure we had everything we were responsible for covering. I don't know how far from the Iraq border we were, but it was close. My guess is ten miles. No more music was on guard; the guard would be spent in the foxhole looking out.

I dug my foxhole and covered it with my poncho. Unlike Missouri, digging in the desert is easy. I grew up digging on the beach; it's the same deal. Typically, we knew no one, just the engineers, but I could see an artillery division next to us.

The company commander said, "Over the next week or so, we will put our plan into motion. Once we reach our designated area, we will load our tracks, and ACE Earthmovers—a new armored bulldozer— and even the Humvees, deuce, and a half up with all our anti-tank mines. You can see where I'm pointing here on this map. The

Revolutionary Guard will come from this northern area, heading directly towards this location. They will come down through this pass here. We must work fast. We won't take the time to bury them. We want the enemy to see them. We want them to turn their forces into this canyon where our forces will wait. This is called a kill zone, or more accurately, a turkey shoot."

He moved his finger around the map. "We will not screw the rod into each mine either. Some of you soldiers will take the rods and stick them in the sand, rows and rows of rods attached to nothing. We want them to think we buried those mines. This will save us time. We will break everyone off and tell them their assignments. Lift with your legs; the mines are heavy. One person places the mine, and the man behind him turns the dial on it and arms it. We will put a mine every three feet. This is called a deterrent. We need them to turn. If they don't, we're screwed. These are the baddest boys in their army. We have another plan, but that is a last resort. I don't think we'll need to resort to it, but we know a backup plan is in place. Do your job, and do it fast. We need to make them regret ever being born. Now, break off to your units and get to work!"

We spent the next few days loading and unloading mines from the back of our track. Practice makes perfect. Every vehicle lined up in formation. We were about ten feet apart and slowly moved forward, with one person sitting on the back of the track. Bunk would hand me the mine, I would place it, and Johnson would come in behind us and arm it.

Bunk complained the entire time. "My goddamn bunion is acting up, but I need a warm water foot bath. Hey, Joe, do you need me to hand them to you? I can stack them, and you grab them."

"Fuck your bunion! Fuck your foot. No, fuck that. Fuck both your feet!" I screamed.

"They are timing us. If we don't speed up, we will never get a break. You are sitting. How much could it hurt? Bunions don't act up!" I replied.

We would not arm it but would go through the motions. We laid the mines down for about half a mile; then, we would reverse course, pick them all up, and load them back onto the track again. Repeat all day. We were timed. The following week, after all our mine deployment training, we moved to a rallying point near the border of Iraq—days filled with more gas mask training. We were getting closer and heard the air war was nearing its end. It might be absolute in the next few weeks or months.

I was told the Iraqis did not stand a chance if we gassed them; our gas would eat through their mask and kill them painfully in seconds. The Iraqis still had not pulled out of Kuwait. Instead, we started getting reports of the horrors being leveled on Kuwait citizens—rape and murder of women and children. Saddam set oil fields ablaze. We could not see this on the horizon yet, but we were getting closer. We drove six to eight hours per day. It was dusty and loud. Most days, I would stick my head out of the open hatch just behind Sgt. Steward. I would try to nap near the engine access panel on a box filled with explosives. The constant rocking of the track would shake me awake.

We still didn't have any ammunition. The higher-ups did not trust us. We could all read maps, and we knew where we were headed. We were going to be a stone's throw from the Iraqi border. I wanted bullets. We all did. I usually spend my entire guard shift in my sleeping bag. I would hop around if I needed to get anywhere. As we got closer to the border, I began to take guard much more seriously. The days bled into each other. It had been roughly six weeks since the air war began.

The company commander called a meeting. "I want you to hear this from me. I don't want you to listen to rumors or spread inaccurate information. We bombed the bejeezus out of the enemy, and they still haven't decided to give up. We are going to cross the border in a day or two. We still have work to do. War is upon us. The air campaign is nearing its conclusion. I don't know exactly when, but it will be very soon.

"The following steps are as follows: we will stow all unnecessary equipment. We will leave it up to your squad leaders to assign your duties. We have trained, and you are ready. I know there is much fear, fear of the unknown. I also have that fear. I have a family. I want to see them again. But we will all leave if we work together, watch each other's backs, and listen to our squad leaders. The military is unique because we learn each other's jobs; if one man falls, the man next to him takes up his job and finishes the task. For example, we will be passing out ammunition after this meeting. Magazines are loaded and will be handed out. You will sign out your ammunition. You will not discharge your weapon unless instructed. Am I understood?"

"Hoo-rah!" the entire group yelled in unison.

"The armorer has everything ready. After this meeting, we will break up, and you will get your ammunition. Guard duty will be doubled. We're close to the border. You will be armed. We will rely on our training. Ensure you are familiar with the daily challenge and its corresponding response. If you don't, this can get you killed. You can engage if you challenge a figure in the dark, and they don't respond. Get your ammo, get your assignments for tomorrow, and get some sleep. Once we cross, we have no idea when you will be able to sleep again. It would be best if you stayed alert. Good luck to all of you, and I will see you on the other side."

I could not obtain ammunition that night; they had been overrun and had run out of packed magazines. The following day, Caldwell and I were called to duty early. We loaded up the back of a Humvee with pickets, our lieutenant behind the wheel. We do all the work; he sits in the Humvee. These pickets were typically used to hang razor wire, but they are now used differently. With our picket pounders in hand, we set out to mark the route to the Iraq border. The picket pounders were heavy, fifteen-pound hollow tubes made of heavy metal, with one end closed and the other open, complete with handles on each side. As we lifted the 17-pound tool over our heads to pound the pickets into the ground, it felt like I was raising a sack of bricks over my head. I had arms like pipe cleaners. My shoulders ached with every lift, and I couldn't help but envy those with more muscles.

We marched on, pounding a picket into the ground every few hundred yards, marking our path with the flickering glow of chem lights that we had snapped and shaken from a box. Though the lights produced no visible color, they were chem lights that could only be seen through night-vision goggles. It was a simple but exhausting process, and by the time we reached the border, my arms felt like they had turned into Jell-O.

The border was marked by a twenty-foot-tall dirt berm stretching as far as the eye could see in both directions. It was an imposing structure, unbreeched thus far, and one that we would soon cross into Iraqi territory. I couldn't resist the urge to relieve myself on the other side of the border, but a sound in the distance caught my attention as I did.

It was a giant Hind-D helicopter looming closer and closer. The chopper was a prominent Russian aircraft often featured in action movies; this helicopter constantly tried to kill Rambo. It had large arms on each side, bristling with guns and missiles.

It was headed straight for us. The sight of the chopper sent a shiver down my spine, and I could feel my heart pounding. For a moment, I wondered if we would make it out alive.

"Ummm, Murgia, get into the Humvee!" the lieutenant yelled from the other side of the dirt berm.

The chopper was getting closer, the sound of the blades getting louder. I finished up my epic urination and ran over the berm. Caldwell and the lieutenant were already leaving without me! They stopped, and I got in.

It was now right on us. We sped back towards our unit. I saw tanks and a couple of Apache helicopters coming toward us. An A10 flew overhead, but the aircraft did not engage. Nevertheless, the roar of those engines was unmistakable.

The helicopter crossed the border into Saudi Arabia. We had no idea what was happening. When it finally landed, they waved a white flag.

That was the first of many surrendering soldiers we would encounter. Most dropped their equipment, including their boots, where they stood and walked home.

We got back to our unit. We were just in time to receive our live ammunition. That came with so many warnings. It was as if they were afraid to give it to us. They didn't trust us, and probably for a good reason. All our hackles were up. I now knew why they waited until the last minute to give us ammunition. The guard was now different. You had to walk the entire perimeter around your track vehicle.

I teamed up with Gus. He was in another squad. We were both low-ranked, so we worked together a lot. You couldn't be a lower rank. Being good friends made it fun. We got the early guard, so it was early evening, and we would get a nice long sleep. We memorized our challenge and responses right before guard. However, we also loaded up on coffee and hot chicken soup.

For example, I'd challenge a soldier walking toward me and say, "Thunder."

Their response should be, "Lightning." Of course, you would probably be put down if it was anything else.

We were too close and heavily armed now to walk around, not following the rules. I was on guard as I walked through our camp with Gus. We were at our post and saw a figure approaching us. It was a pitch-black night, with clouds covering the Milky Way.

We challenged, "Thunder."

This figure said nothing and kept walking towards us. It was too dark to see his shape. He was a hundred feet away at this point. We challenged again, but nothing kept walking toward us. My heart is now beginning to pound. He was walking in almost complete blackness. He was just a silhouette; it was not a smart move. We all had itchy trigger fingers.

"Thunder!" Gus repeated.

The figure continued saying nothing and kept coming toward us.

I looked at Gus, and he looked back at me. We were both locked and loaded. The unmistakable sound woke the figure up. He began shouting and holding his hands in the air. It was Clapton, our medic.

Clapton began to panic. "Jesus, lightning, lightning. It's me, Clapton."

"Hey, dipshit," Gus said, "in case you can't tell, it is pitch-black out, and we can't see who you are. You're black, making it even more difficult. We can't even see you when you smile! Do you want to fucking get shot? It would be best if you spoke up when challenged. Did you not hear the captain?"

"All right, all right, you guys, sorry."

I was ready to light him up. Instead, Martinez walked up to us with a few days of growth on his face. He was also dirty for some reason.

"You look Iraqi with your beard! You may not make it, Martinez." I said, pointing at his face.

"Fuck you, guys!" He shouted back.

"Praise Allah, motherfucker, are you working for the other team? I have a prayer rug you can use," I teased.

"I will fuck you guys up," Martinez snapped.

"Remember not to respond to the challenge in Muslim, not Spanish or English! Motherfucker!"

"Yeah, I'm hitting the rack." He said, "See you guys in the morning. It's a big night tomorrow. Get some sleep."

"Have a good night, Sergeant."

"You too," he replied.

As the guard shift ended, I returned to the track, scanning for signs of movement.

But a loud explosion shattered the eerie silence of the night. Barely five hundred yards away. Is Michael Bay attacking us? Maybe? Most likely, they are Iraqis.

My heart raced as more explosions followed, each smaller than the last. The chaos and confusion were palpable as people scrambled toward their foxholes, and I followed suit. We were under attack and dangerously close to the border.

I spotted a burning howitzer through the chaos, its shells exploding in all directions as they cooked off. The concussion in my chest was almost unbearable. I felt it in my marrow. I crouched in my foxhole, watching as groups of men frantically tried to extinguish the blaze in the diesel truck area.

We hunkered down in our foxholes for an eternity, waiting for the danger to pass. And when it finally did, we were told that it was a false alarm—this was all due to an electrical issue, and the artillery shells in the back had started cooking off and exploding.

But the fear and unease lingered, and I returned to my foxhole, eager to sleep. However, as luck would have it, a heavy rainstorm arrived, and my hooch was not ready. Water seeped into my foxhole, and I used my trench tool to scoop out some of it. I made the best of the situation, adjusting the poncho and spiking it into the ground to provide some cover, but it was no use, and I slept half-submerged in a one-foot-deep puddle.

The next day, I woke up feeling miserable and cold, surprised that I didn't get hypothermia. But there was no time to dwell on it—we were set to invade Iraq that night.

So, I hung my half-wet sleeping bag on the track vehicle to dry in the sun and jumped in the Humvee with the lieutenant, driving the route to the border again. We ran the assigned errand that morning. On our way back from the errand I can't remember, we spotted a giant tent with a lot of bustling activity, clearly American. We had no set time to return, so I asked the lieutenant if we could go over and inspect this tent. I've never seen a tent this prominent, possibly four stories high and as long as a football field - an American football field, 100 yards long. I'm an American; we compare everything to the length of a football field. In my brief time in Georgia, everything seems to be a stone's throw away. They also did a lot of fixin' and reckoning.

As I approached the massive tent, it loomed over me, its enormity intimidating. The sides were adorned with the scars of the elements, and the creases on the fabric indicated that it had been there for some time. The entrance was draped with a thick canvas sheet, and as I pulled it aside, a rush of stale air escaped. Inside the tent, the atmosphere was thick with the smell of oil and machinery. I saw rows of metal shelves with everything from canned food to batteries. The aisles were filled with crates labeled with a destination, ready to be loaded onto trucks and taken to their respective units.

In the corner of the tent, a group of mechanics was working on a Humvee, tools rattling in the background. A few soldiers sat around a table, playing cards and sipping coffee. I could see the relief on their faces as they took a break from the rigors of the war. I couldn't help but feel a sense of awe as I walked around the tent. This was the operation's logistical backbone; we would be stranded in the desert without it. It was a testament to the efficiency and determination of the men and women who ran this place.

A guy walked over to me. I thought I was about to get the riot act read to me when he said, "Hey, what unit are you with?"

"3rd Engineers."

He looked at his clipboard. "Cool. Are you here to pick up your donated supplies?"

"Donated?" I asked.

I looked at the boxes in the tents—cans of chili, M&M's, Vienna sausages, juice boxes, SpaghettiOs, and many others I can't recall. The boxes were stacked on top of the tent. It was like being in Wonka's factory, but it was all corporate, processed foodstuffs, not real food.

There was no Vienna sausage fountain, as I had dreamed. No army of Oompa Loompas working the tent. Boxes of processed food as far as the eye can see, or at least to the end of the enormous tent.

"Nobody from your unit has picked any of this up. Grab what you want."

I looked at the lieutenant and said we needed to get a truck or an ACE vehicle and pack it up. So, I grabbed a few boxes of things and loaded them on the Humvee.

We drove over, and I found an ACE Earthmover driver. I told him about his new mission. We went to the tent, liberating as much candy and Vienna sausages as possible. He pushed the dozer blade as far forward as possible, creating a wide-open space to load the goodies. The driver and I filled the ACE up to the brim.

I must have eaten my weight in M&M's while loading everything up. Then, we hopped back in the ACE, drove to each squad, had them take what they needed, and loaded it on the track. We were heroes—bringers of candy and tiny sausages. In medieval times, songs would be written about us. We were Santa handing out diabetes to all of our friends.

Our unit commanders purposely didn't tell us about the supplies offered; they didn't want us to get too spoiled. It was okay for some people to profit from their troops, but not to provide us with supplies that might make us a tad happier. They were mad that I told everyone about the supply tent.

As we sat at the border of Iraq, our stomachs full from our last meal, we were summoned to form a line, ready to board the track vehicle. A hush fell over the group as we were each handed a pack of pills, the back shimmering in the light like a tiny silver beacon in the desert. They were minuscule, round dots that looked so small they could easily get lost in the creases of your palm. Our commanding officer sternly ordered us to take one immediately and another every four hours, warning that they would check to ensure we had swallowed them.

The mystery of the pills heightened our curiosity, as there was no branding or writing on the back of the package, unlike the medications we were accustomed to taking. I couldn't help but feel a sense of apprehension as I pretended to swallow the pill, my mind racing with questions. I know how the government likes to use its troops as test subjects without their knowledge, and I remember Agent Orange, so I spit the pill out after pretending to swallow. What was this medication, and why were we taking it? Was it meant to protect us? The lack of information only fueled my curiosity.

"What is it?" I asked.

I was met with, "Just take it!"

"No, I'm not taking anything," I said. "Just handed it to me without writing. Does anyone remember Agent Orange? I do. I'm not swallowing shit."

At that point, everyone who was about to take it stopped and started questioning. I got the death stare from Sgt. Martinez, who handed the pills out to us.

"If we get gassed," he said, "this will give you one extra second to put on your mask." You will thank me if we ever get gassed."

"No thanks, Agent Orange. I don't want to survive a gas attack."

"Take the pills—one every four hours. "Don't forget," Martinez demanded.

"I wouldn't take that shit if I were you guys. Agent Orange! Agent Orange! Don't be their test monkey!"

Atropine was issued as part of our nuclear, biological, and chemical kit._Atropine is a tropane alkaloid and anticholinergic medication used to treat certain types of nerve agent poisoning, pesticide poisoning, slow heart rate, and decreased saliva production during surgery. It is typically given intravenously or by injection into a muscle. Our Atropine had an auto-injector that would NASCAR that drug into your system.

The Iraqis also had a nuclear, biological kit. It wasn't as advanced as the US version. They had a more street-level kit. If they were gassed, they had a mask, and instead of Atropine, they had glass vials filled with cocaine. This was the rumor, so quite a few troops were trying to find those kits. I never saw one. I can't confirm if this is true, but it makes sense.

After the war, we were having Atropine wars and shooting it at each other through a cardboard box, and when you got it on your tongue, you'd be gacked out for ten minutes. Even a light dusting of the drug was super intense.

My heart would beat, and it felt like it was hitting my rib cage. The gall of the government to think I was going to take their word and ingest that poison into my body. I had already eaten so much shit smoke, and enough damage had been done.

So, I threw the rest of the pack of pills into the desert. This did not stop one soldier, Davies, who was incredibly nervous and started popping these pills every few minutes instead of every four hours; he eventually passed out and was convulsing and foaming at the mouth. We had a medic nearby, but I'll never forget the helplessness I felt watching him go like a shaky, rabid dog.

I knew those pills had no FDA approval, and it was on brand for the good ole U.S. of A to test drugs on their soldiers, only to fight after the war to not take care of the very people who defend their freedom. You are not a person to the government, but a number who signed away your health and rights. I never got the Gulf War syndrome later in life since I never took those pills (*Pyridostigmine Bromide*, an anti-nerve agent). Some of my fellow soldiers didn't listen and had many health problems after the war. It was determined that the pills, mixed with our issued mosquito repellent, and other factors, caused this illness later in a lot of troops who served in this combat environment.

CHAPTER 8

*I*raq, *here we come.* We loaded into the back of the track vehicle. Bunk and Johnson settled into their spots. There was one more person with us on the track, and to this day, I can't remember who it was; they are a blank spot in my memory, not just for me but for Johnson, too. We have discussed who it might be, but we always come up blank. Whoever it was, the invisible man contributed nothing else to my story. Just know that someone else was in the back of the track with us; they did nothing as far as I can remember. The prospect of a long drive ahead only added to the tension building within me. As I climbed onto the aluminum structure, I couldn't help but feel uneasy. This wasn't a tank; anything more significant than small arms fire could easily penetrate the "armored" walls.

Despite my worries, we pressed on, with our driver at the helm, navigating the treacherous terrain ahead of us. The hatch was open, exposing us to the elements and any potential threats that might come our way. To my right, Sgt. Steward was operating the fifty-caliber machine gun—a formidable weapon, but one that only added to my sense of unease.

Our path had been carefully planned, with my lieutenant and Caldwell leading. We crossed the dirt berm separating Saudi Arabia and Iraq, and the low rumble of our convoy was a constant reminder of the danger ahead. As we crossed the border on February 24, 1991, we knew the journey ahead would be long and grueling.

And it was. We drove all night and well into the following day with nothing but the endless supply of dust and dirt for company. The constant barrage of exhaust fumes only added to the misery, and I couldn't help but feel that this, combined with the suffocating MOPP suit, had taken several years off my future.

In the following months, I would continue to feel the effects of our arduous journey. Chunks of charcoal would inexplicably fall from my skin and pores, a constant reminder of the toll the war had taken on my body.

As we pressed northward, the 24th Infantry's heavy armor division leading the way, we encountered sporadic resistance from demoralized Iraqi troops. But it was nothing major, just some stray gunfire here and there. By noon on the twenty-sixth, we had traveled two hundred miles north and had successfully blocked a major Iraqi supply route and possible avenue of retreat.

For a day or two, the Middle Eastern front was quiet. But then, on the horizon, we spotted something that would forever change my perspective on the war. Little dots, barely visible in the distance, slowly came into focus as we drew nearer. Families with children, men carrying all the belongings they had in the world, which, to my surprise, didn't look like much. We handed out food as we approached and threw boxes of MREs to the children.

It was my first experience with refugees and the human cost of war. The realization hit me hard. The idea of people being displaced had never crossed my mind. Yet, as I looked into their eyes and saw the toll it was taking on them firsthand, I couldn't help but feel for these people.

Though I had recently graduated from a dehumanization camp and had been trained to view the enemy as less than human, some of my humanity had survived.

As we moved forward, the seemingly never-ending line of refugees stretched like a sad and endless procession. The scorching sun was beginning to dip below the horizon, casting long shadows across the dusty road. Our formation moved steadily onward, the low hum of the engine providing a constant backdrop to the haunting silence that had befallen us.

The sky began to darken. We had been driving for a day and a half, and the endless monotony of the road had taken its toll. My eyes were heavy, my head drooping, and I fought to close them again, knowing that the images that would greet me would be just as haunting as before. Then, finally, it was late afternoon, and I heard the words that sank my heart.

"We got two guys with AK-47s up here. No engagement yet."

Dusk created an orange-pink hue across the sky; my heart raced with fear and excitement. The beauty of the day seemed to contrast with the impending violence ahead.

Then, the unmistakable sound of gunfire broke the silence, sending chills down my spine. It was a sound I had trained for, but the reality was far more terrifying than any simulation.

First, the sporadic shots would stop abruptly, followed by an eerie silence that was even more unsettling. Finally, the command to move into an attack formation snapped me back to reality.

Bradley's and M1 Abrams tanks led the way while we, as engineers, hung around to support the infantry. Our job was to lay mines, but that would only happen once we arrived at our designated area, which we might not reach until the next day.

Small arms fire erupted from all directions, and I struggled to distinguish between our soldiers and the enemy. The sound echoed in my ears, and my mind frantically logged every detail—the sound, the direction, the distance. Now, there was a distinct ringing in my ears. My senses were heightened, my eyes scanning the horizon for any movement.

The rush was tangible, and I felt every inch coursing through my veins. My vision sharpened, and my hearing became more acute. My head swiveled at the slightest sound; my body was poised to react to any threat. But, instead, the never-ending expanse of beige sand that stretched out before us felt more ominous than ever, and the gunfire made it all too real.

However, my squad mates remained asleep, oblivious to the danger surrounding us. I kicked them awake, desperate for them to feel the same urgency consuming me. It took more kicks than it should have.

"I'm up, I'm up," Johnson groaned.

"Shit is going down. Get the fuck up," I snapped.

Gunfire from all directions. The rapid pops of the AK-47s mingled with the more resounding blasts of the M16s, creating a discordant symphony of war. I could feel the weight of my weapon in my hand and the urge to shoot at anything. It was almost overwhelming. However, I knew I had to remain calm and focused, waiting for the right moment to act.

I looked over my shoulder, scanning the darkness behind us for any signs of danger. The only person I could see was Bunk, who had finally managed to shake off his narcolepsy and looked alert and focused.

The tracers streaked across the sky like fiery comets, illuminating the night and revealing the enemy's movements. We took cover behind our vehicles, listening to bullets whizzing overhead.

I could feel the heat of the desert on my face and the sand crunching beneath my boots. As we waited for the fighting to subside, my mind wandered to Bunk and his bizarre sleeping habits. His constant snoring and talking in his sleep had been a source of annoyance, but now I couldn't help but feel a sense of camaraderie toward him. We were in this together, fighting for our lives in a foreign land. And as the night wore on, I knew we would need to rely on each other more than ever.

"Joe, have you got hot chow yet?" Sgt. Steward asked.

"Not yet," I replied.

"All right, when we stop, get back to that Humvee, and they'll take you back and fill you up."

I jumped off the track immediately and ran over to the Humvee. I could hear sporadic gunfire in front of me. We were on schedule, so we had to eat fast. We drove back a few miles from the front, and a full kitchen ran. I got in the queue and received a full plate of hot lasagna, cornbread, and a couple of sausages.

In a few days, I'd speak to some Iraqi soldiers who thought they were on the border but were two days in and were surviving on a spoonful of rice three times per day and water. They never stood a chance.

I was trying to savor that meal, knowing I would be moved up near the front soon enough. Army food is typically not described as a Michelin dining experience, but I'll tell you, it was the best lasagna I've ever eaten. There is nothing better than hot food when you aren't expecting it. It went down quickly. I didn't even stop to taste it, like always.

Apache helicopters swooped in, firing at unseen targets with a thunderous roar. The AK-47's distinct sound was in the air, setting my nerves on edge. Squeaky tracks could be heard moving forward, and armored personnel vehicles and Bradleys were ready for battle. The backdrop to our meal was the loud sound of a .50-caliber fire, the rhythm of the war zone. A cloud of dust hung over the entire area, obscuring our view, but we could see the explosions in the distance as the front line moved away from us. Fear was palpable, but relief tempered it that the waiting was over, and the fight had begun.

A new set of emotions emerged after returning to my unit: dopamine and serotonin surges, testosterone, adrenaline, and sadness. I felt my mortality, and the feeling was different from fear.

Instead, it was a sense of dread offset by the natural chemicals coursing through my veins. I didn't think I would die, but I couldn't shake the feeling of death's skeletal hand on my shoulder. I knew I was young and invincible, but that feeling wouldn't last forever. As I arrived back at my squad, I saw that we had already set up a perimeter, circling the wagons.

My friends were stationed at their .50 calibers, their eyes locked and loaded, ready for anything coming our way. We were a team, prepared to face whatever the enemy had in store for us. I saw the ACE vehicle we loaded up with goodies, and I grabbed chili, spaghettiOs, and as many Vienna sausages as a human could carry.

One of the first things that became apparent when we settled into our perimeter was the presence of portable stoves, and I saw a large coffee can filled with water at a rolling boil. I said armies march on food in their stomachs, but coffee is a close second. So, I dumped a few cans of chili into the boiling water and distributed the other food to my friends.

Sgt. Steward greeted me and immediately said I needed to get some shut-eye. How? They would wake me up in a while for guard duty. I pulled out a cot, set it up, and grabbed my poncho. I pushed it up against the track, facing the inside of the perimeter. I lay my rifle by my side, pulled the poncho over my head, and closed my eyes. I heard gunfire, explosions, and the sound of tracked vehicles moving past us. Less than ten feet away, someone started firing their .50 caliber at something or someone very unlucky. You could hear the howitzers shelling the place we drove to a few miles away in the morning. You could feel the ground shake when they were firing, even a few miles away.

How the fuck am I supposed to sleep? I realized I wasn't.

I saw a Bradley vehicle firing a .50-caliber machine gun. The barrel was white-hot. They stopped firing, and I saw them remove the heat-resistant glove, start unscrewing the barrel to replace it with a new one, and keep firing. They were shooting at something behind me.

There were very few lulls, but now and then, you would hear distant screaming. A fearful scream differs from a scream derived from pain and agony. You can feel those screams.

You always remember them, and the smell is something I wish I could forget: diesel mixed with the scent of burning hair. There was blood in the air—every breath you took stung—gunpowder and smoke from all the vehicles. I also saw others setting up gas alarms around the perimeter.

I was so overwhelmed with what was happening around me that I completely forgot the other possibility of sarin gas. I did not want to go out that way. I didn't think I would like to survive if we were gassed. I felt for and clutched the gas mask on my hip after the reminder of the gas alarms. Each squad had its own. The alarms never went off when I was around, but years later, in a very blunt letter, I was told we had been exposed to sarin, thanks to the wind.

I don't think I ever experienced any symptoms, but we *were* exposed.

After a few minutes, we loaded back on the track and drove further down. Tanks and trucks in the distance burned brightly, lighting the darkness with an eerie orange glow that pierced the blackness. Screams in the night. Bodies are everywhere on the ground. Pools of viscera soaked into the desert floor. Blood and guts everywhere you looked. We passed burning people on our right and left. Vehicles were now nothing more than twisted and mangled piles of metal, with bodies slumped over in their seats. The tank fire had left its mark on the desert landscape, leaving plumes of smoke and dirt in the air. Scorched sand surrounded us. Bodies are hanging out of the cabs of their vehicles. They had died trying to escape. We were gasping for air, our lungs burning from the thick dust clouds and the pungent stench of diesel fuel.

Our unit met up with a Marine unit in Humvees. I have no recollection of what unit they were part of. Some soldiers were on foot. I have no idea what their unit was, and I'm sure they mentioned it to us. My marijuana-addled brain can't remember details like that anymore. They jumped off the Humvee and walked towards us, slinging their weapons on their shoulders. Some of the men were getting their trenching tools out of their rucksacks.

"You guys' Army?"

"Yes, are you?" I replied.

"No, Marines," he replied.

They must be infantry, walking beside mechanized vehicles and Humvees. I saw Corporal Beck behind us. I waved, and he waved back. He was sitting behind his 50-cal.

We continued to drive through this post-apocalyptic sight. We saw some Marines pulling bodies from the back of a truck. Other Marines pull a body out of the cab of a smoldering truck. It wasn't just military vehicles burning; regular GMC vans, painted in desert camouflage, also existed. The Marines were walking around and taking the heads of the newly deceased Iraqis. They were holding their hair and hacking at their necks using their trench tools. It wasn't smooth, like using a samurai sword. It took several hits to get through all that muscle and bone.

You could see the spinal cord. It's fucking repugnant.

"What the fuck!" I yelled.

All I got back was "Souvenirs."

They placed the heads on the front of the Humvee, much like a hood ornament. Blood dripped down the front of the vehicle. I had no idea how they would keep their heads from falling off, but I wasn't sure they thought that far ahead. Several Marines followed suit and started hacking off the heads of dead soldiers.

We slowly separated; the Marines went their way, and we went ours. The fighting continued. Sgt. Steward would hand me the night-vision goggles so I could see what was happening ahead of us. I heard rifles locking and loading. I rushed over, and an Iraqi man was walking towards our perimeter. A couple of soldiers locked and loaded and screamed at him to put his hands up. He walked towards them, his arms crossed, like Dracula in his coffin.

I was preparing myself mentally, knowing that I was going to watch someone get shot and die, and I was no more than fifteen feet away from him. I could still see his face and the fear etched into it. As he got close, he raised his hands above his head, and his entire throat and chest were gaping open wounds as he did. His innards poured out of him, and he dropped to his knees.

The screaming changed, prompting a medic to help. He lived for a few more minutes. I can't get the gurgling sound out of my mind, even all these years later—the struggle to breathe, the humanity the government wanted me to erase, and mow the enemy down. I was faced with a human being unable to gasp for air; he was sucking his blood into his lungs.

I still see his face sometimes, but not as much as I used to. This Iraqi man would visit me when I had too much to drink, but I drank so I wouldn't see his face, so the joke was on me.

We were moving in the morning. Sgt. Steward could see sleeping was not going to happen, so he started going over what was happening and what we would do the next day. He urged us to get sleep over and over again, but it was not physically possible for me.

I'd be ready the following day, even with no sleep. You never forget that your life is at stake. It was late, probably around 2 a.m., and I was told I was on guard duty. I don't think anyone around us got any sleep that night. I took guard duty very seriously then.

There was no music, no headphones. I suppose I went there to fight a war. I sat on the 50-cal, watching the flashes of explosions on the horizon. Eventually, I jumped down and walked our perimeter. I wandered around the track and saw the portable stove with a coffee can of boiling water; cans of very hot chili were inside. The air was cold, and hot chili would hit the spot.

I ate my chili sitting behind the 50-cal. I lowered the seat to give someone less of a chance of shooting me while I enjoyed my chow. That might have been the best meal I've ever eaten. I was worried it might be my last meal.

I put the night-vision goggles on. I saw Iraqis in the distance, hiding behind burning trucks and firing in every direction. Apache helicopters swooped down, shooting missiles at tanks. Gunfire was all around me, and it never died down. The tanks in the distance exploded, some turning over in the air and landing on their turrets.

I saw a U.S. soldier on a Bradley shooting a .50 caliber at a line of Iraqi troops. They were about two hundred yards away and were trying to reach the top of a hill so they could dig in behind it. One soldier tried to run to the top of the hill. The 50-caliber hit him, and the center of his body was eaten away by gunfire ripping through his torso. He slumped forward. We drove towards this area later the following day, and he was blown almost in half, his spine exposed to the morning air.

I watched in horror as trucks careened across the highway to their right, troops scrambling to avoid the hail of bullets that rained upon them. This road is designated Highway 8. It runs from Basra to Baghdad. However, it would be known by a new name: The Highway of Death, after the next few days.

CHAPTER 9

As I watched the sunrise on that chilly February morning, the silence was eerie, interrupted only by the distant thud of howitzers and sporadic gunfire. I felt numb, my thoughts jumbled, and my senses heightened. It was quiet and made me wish for the comforting sound of a cassette playing, but the rules of war prohibited such a luxury. I had to stay alert, aware of my surroundings, and cautious of any possible danger.

The day passed quickly as we made our way north and turned west towards Kuwait, the landscape dotted with refugees fleeing the warzone. We were all aware of the looming threat of the Republican Guard of Iraq, the boss-level fight we had been warned about. The regular Iraqi Army was just NPC characters to us, obstacles to overcome on our way to the big bad. We were told the Republican Guard was well-funded, well-trained, and dug in, with loyal soldiers who had housing and even bonus pay for their loyalty. However, we soon discovered that it was all just propaganda. They still used our old equipment to fight us; it didn't matter how well-trained we were if our equipment was borderline useless. I felt for them.

The sheer power of American technology was awe-inspiring, even if the cost was a bitter pill to swallow. Watching an A10 Thunderbolt strafe tanks with its Vulcan cannon was a sight to behold, the deafening sound of six thousand rounds per minute echoing through the air. The A10 was a marvel of engineering, from the sound of its engines to the precision of its weapons. It was no wonder the Republican Guard rarely had any success against it. Meanwhile, the Iraqi Army still used World War II-era equipment, ranging from rifles to pineapple grenades.

Their 1970s-era GMC vans were no match for our high-tech vehicles. They might as well have been driving a *Toyota Prius*. So, I was amazed to see a track vehicle run over an anti-tank mine and come out relatively unscathed. I had expected something like that to decimate our vehicles, but we rolled on undeterred instead.

Looking back, it's hard to believe I survived such a brutal and unforgiving war. But at that moment, I could only watch the sunrise, take a deep breath, and prepare for whatever lay ahead. Still, the only real damage was to the track, which, within five minutes of running over the mine, was already being worked on and replaced by the crew. I remember thinking how fucked the Iraqis were at that point, and I felt something I rarely feel in the world: pride. I had at least one hot meal daily, if not two or three. Watching these guys jump out of the back of the track and immediately get to work was inspiring. There was no delay. Each man had a job, and they got to work. I've changed tank tread before, which is challenging work, but watching these guys gave me hope. Finally, finally, I might make it out of there.

I worried about driving over a mine because of the two wooden boxes filled with C4 plastique explosive and leaky dynamite. If that went off, I would be a pink mist. We also had some Bangalore torpedoes, which were three feet long and packed with explosives. They could be connected end to end and pushed through a minefield, allowing you to blow a drivable lane through one. I'm not sure how many we had, but each one had roughly up to nine pounds of explosives in each torpedo. The mines could be detonated by simply driving over them, like an idiot, but you could also attach a small rod, and if a vehicle pushed that rod more than thirty degrees, it would set the mine off as well. We had a ton of rods that we would stick in the sand to create the illusion of many more planted mines.

Unfortunately, all that training would not pay off because the plan changed, since, within the first couple of days of the ground war, we wiped out most of the Republican Guard. There had been seven motorized infantry divisions, and by day two of the fight, five had been erased from the Earth.

They had a paratrooper division and some support divisions, but they were held back. Of course, the motorized infantry was the primary concern, but the news spread quickly that we had already gotten them on the ropes.

The late-afternoon sun cast a warm glow across the landscape as our troop established a perimeter once again. Sgt. Steward was the one to go to the battalion to receive new marching orders. The plan to lay the mines was no longer necessary, and things were changing. It was a moment of relief and a reminder that the situation was constantly evolving. Highway 8 was perched atop a hill and looked like any other multi-lane road. It could be Arizona if I didn't know any better. As the formation grew tighter, more Bradley-fighting vehicles appeared ahead of us. The highway was impressive, featuring overpasses and intersecting roads that ran north and south at several points. However, the road on our left was so broad that it felt almost endless.

We began to slow down the formation. It was very dark. We slowed, but some of the Bradley fighting vehicles continued moving forward. Meanwhile, some M1 Abrams tanks drove past us while our track stopped.

Over the radio, I heard a voice command to halt. Every vehicle stopped where it was; ahead was just blackness, just a void. The only noise I heard was idling engines. Then, in the darkness, I saw multiple flashes, and seconds later, I heard the booming sound of shells firing at us.

Chaos on the radio. "MOVE! MOVE!"

Bradley vehicles moved left towards the underpass, with multiple explosions just a few hundred yards away.

We backed up at this point. Most engineering vehicles kept backing up until we were out of range. Then, the Bradley sent a round downrange. Helicopters swarmed overhead, raining missiles and gunfire ahead of us. You could feel the thud of the projectiles hitting the ground. Flashes of light downrange. The explosion lit up that void, and hundreds of tanks, vehicles, and tracer rounds approached us.

The concussion of the explosions could be felt in your entire body. It isn't like the movies, where a grenade is thrown and people are launched into the air, with fireballs behind them. Instead, a concussion sucks you in and can suck the air right out of your lungs if you're close enough.

The helicopters were so low that I could reach and touch them as they flew overhead. I heard a Vulcan cannon strafing the Iraqi convoy ahead of us. One tank exploded, and the turret flipped into the air several times before landing about a hundred yards away. The cannons chewed up the desert floor. The orange and yellow flames now lit up the emptiness, and I could see what was happening before us. We were out of their range. We had not engaged with our M16s at all.

The M1 was firing on the move. You could see our shells entering an Iraqi tank and exploding on the other side. Our rounds penetrated their armor like butter. I watched an Iraqi round bounce off the armor of an A1 and explode next to the tank.

On the radio crackled to life again, I heard, "That motherfucker possibly chipped our paint!" before a round was fired that must have set off other rounds because it was a mini mushroom cloud where that Iraqi tank once stood.

Sgt. Steward handed me his night vision and said, "Look at this madness, Joe!"

I put on the night-vision goggles. There was a white blooming effect around all the light. I could barely make out anything. We were still under the road, and a helicopter startled me as it flew above us. It was so low that I felt I could reach up and touch it. This drew my eye up to the overpass. I could see a truck driving down the road toward us.

It was not an American vehicle. The headlights were off, and it was getting closer. Close enough. I took the safety off my M16 and raised my rifle. The truck was traveling at about fifty miles per hour. My fantasy at this point was a one-shot, one-kill, aiming at the driver's windshield.

I was about to alert Sgt. Steward, when I looked up, I saw an Apache helicopter hovering above the truck. I noticed a flash. I got a complete whiteout in the night vision. When the image returned, the car flipped end over end on the road and exploded midair, whiting out my night-vision goggles. However, when the image was restored, the truck was gone and had rolled down the other side of the embankment, away from us.

I handed the night-vision goggles back to Sgt. Steward, because we were now on the move again. We advanced, stopped, and advanced again. We stopped. This happened for the next few hours. Move a few yards, then stop. We moved forward again and paused by a burning truck with unfortunate Iraqis trapped inside, burning alive. Listening to the screams still haunts me.

The smell of the diesel and people burning made me hate myself for signing up and having to experience that horror. Gunpowder also hung thick in the air. I liked the smell, but it began to eat at you after a while.

You could smell blood in the air, soaking in the desert. We also drove over limbs and copper wire, a remnant of firing stinger missiles from the Bradley vehicles, soldiers, or even the Apache helicopters. Again, the wire was everywhere, getting tangled in our tracks. Copper filled the senses. You could taste it on your tongue. It draped over everything. Blood on the tracks, seeing bodies run over repeatedly by track vehicles until they were indescribable. Mud and blood. I watched several stingers go downrange and hit tanks and trucks downrange. It had a 99% chance of hitting its target if it received a tone or lock. It's an incredible technology that makes me proud to be an American.

It was dark then, but I got a good view the following day. I sat in the back of the track and tried to relax. I used to think I hadn't slept for four days, but I now know that I must have been going in and out of sleep, taking micro-naps, and never feeling fully rested. Caffeine was my friend. I think back, and there are some blank spots from that night. I felt safest sitting down with my back up against the truck driver. It was also close to the engine, and that hum of the motor must have put me out a couple of times. The fumes from the engine probably made me hazy enough to nod off. I heard explosions outside my track; these were close, and my body could feel the concussion.

These were not small-arm rounds. Tanks were firing at us. You can feel the concussion hundreds of yards away. Fear was always there, but I never outwardly showed fear; I knew I would make it. On the contrary, I felt palpable excitement for what would come next.

I am not a risk-taker, but war is exciting, terrifying, and fun, if I'm being honest. That got my heart pumping at full speed. You spent your childhood playing war, and now I was in a war, and tanks were firing at me. It is fun and terrifying simultaneously, but it isn't easy to describe the feeling.

That was how it went for most of the battle. There would be bursts of loud gunfire and explosions, followed by a sudden silence. We drove a bit further, and violence erupted again, then silence. Repeat. It kept you off guard.

It was tough to relax. You could feel your anxiety rise within you, and then you would relax a bit, and it would start again—a hollow feeling in your chest. Your heart felt out of rhythm. This would be an all-too-familiar feeling in my forties. The anxiety of that event that I pushed down deep inside myself would find me again.

We met back up with some other members of our unit. I saw some friends from the 2nd squad. They were walking around and surveying the damage. Everyone seemed a bit disoriented, dirty, and tired. I happily greeted everyone. We started sharing horror stories. I grabbed some coffee. Some soldiers were throwing the bodies in a pile to clear room for our temporary perimeter.

My Native American friend, Gus, was lost. You never know what happens inside someone's head, but I didn't think he was that broken. He was showing us his "souvenirs." I saw him with his bayonet fixed on the top of his rifle. He walked around stabbing dead soldiers lying on the ground, asking, "You dead, bro?" Some of us found it amusing; others found it sad and disturbing. I leaned toward the latter. He saw me, grinned, and ran over.

"Joe!" he shouted. He pulled out a bandana. I saw blood on the outside. He slowly unfolded it.

"You okay, Gus? I think you need some sleep. We both do, if I'm being honest. I can't sleep with never-ending gunfire."

"Nah, I'm good. Check this out!" He opened the bandana, and it was filled with human ears. "Trophies. Good lucking hearing shit in the afterlife!"

Now, this seemed like a war crime, but to be honest, cutting off the ears of corpses felt insignificant in the grand scheme of things. I was worried for my friend, nothing more. We got a signal to load up, and Gus disappeared into the night.

CHAPTER 10

T he sky turned grey, now mixed with some beautiful light blue. The morning was coming. February 27, 1991, began with more coffee and an MRE. We drove under the underpass to the other side of Highway 8, and that was when I saw the aftermath of the truck blown off the road by the Apache helicopter.

The truck had rolled down the hill, the sand blackened by the flames and burning rubber, and came to rest in a chain-link fence. Most of the fence had melted, and the truck's bed had a blackened mess of bodies. One of the men in the truck managed to get out but was dissolved into the fence. He was frozen in time. His hands reached out in front of him, his face locked in terror. His body burned up to his neck, and his face was locked in a scream. I don't know where he was in the truck since it looked impossible to get out of; there wasn't a spot on the truck that wasn't scorched. His face was the only part of him that was not burned, and it was a very grey color. Smoldering vehicles were everywhere, and the smell of burning rubber and diesel choked the air.

After surveying the damage, we retraced our route back through the underpass. Our medics spread over the field, helping injured Iraqis scattered across the desert, giving them water and food, bandaging them up, loading them on trucks, and taking them away.

We were driving east again with the highway on our left, and we began to follow the convoy up the embankment to the paved portion. Turning left to go to the highway, I saw a burned body on the ground. His left leg was broken and at more than a ninety-degree angle. We were covered in blood from head to toe and missing quite a few limbs.

We were less than three feet away, and since it was now a bright morning, I could see him very clearly. His ears were burned off. *He must be dead*, I think, and we continued to drive and pulled up onto the highway when I looked back and saw him waving a white flag.

My first thought was to return and put this person out of their misery before I could say anything. Instead, I saw many of our medics begin working on him. I have no idea what became of him.

Lifeless soldiers and refugees are on the highway. People huddled together, eating and shaken up by the last few hours. We told them to clear the road as more of our unit followed us onto the highway. Smoldering trucks were all over the road; we had to weave through them, and some had to be physically pushed out of the way.

As we waited for a vehicle to be moved, we parked next to a truck facing in the opposite direction. I looked at the driver's side. The windshield had a huge hole. There was a man in the passenger seat, his head leaning against the window, with a lot of brain matter oozing out of his head. He was gone, but something was very wrong. I stared at him, trying to work through it.

That was when I heard Davis from behind me say, "Oh shit, his head is missing!"

At that moment, I blinked, and his head disappeared. My brain protected me from trauma by adding a head back where one was missing. I understand I was in shock. His head had been removed by whatever made the hole in the windshield.

I had to sit down. I felt sick and unsteady. Then, the track began to move. We moved about a mile on the road. A line of different types of vehicles lined the road, some of which were smoldering. A truck loaded with ammunition was on fire and cooking off rounds in every direction.

We got closer and closer to it. Finally, we were no more than thirty feet away, and Bunk and I sat at the back of the track. We parked ten feet from a truck, cooking off rounds. They were popping and whizzing in every direction.

I started to turn and ask Sgt. Steward to move us away when a round hit the top of the track, chipped the paint, and a round whizzed past my ear. It was so close that it brushed against my ear; I felt it move and heard the whizzing sound.

Bunk saw it happen. "Oh shit, that was close!"

"Hey, can we move this fucking track out of the range of these bullets cooking off?" I yelled.

I cannot emphasize enough how this one near-miss changed me entirely as a person. That bullet whizzing by my ear killed the introvert inside of me. I felt I had not fully lived life and wanted to live it fully. I will never forget that sound the bullet made cutting through the air. That was the moment when I made a deal with myself. *If I survive this, I will savor every moment. I will try to fuck every girl I can, savor every meal, be a better friend, and stop being afraid.* The last one was more about being too much of an introvert. I wasn't going to hold my tongue from that point on. I kept most of those promises to myself... for about a year.

We were finally out of danger from the killer truck cooking off rounds in every direction. More and more refugees were coming toward us, and we were now following our convoy toward this village in the distance. It wasn't enormous, and some buildings were partially standing. They were made from mud and straw, similar to adobe houses in the American Southwest.

All the structures were beige. Leaflets were scattered on the ground, and military equipment was piled up. It was made of fifteen to twenty buildings. We pulled up, and the back of the track opened. I saw some members of my squad. Sgt. Martinez ran up to me. He was so happy to see me. He asked me how I was and if I had seen this and that.

He wanted to hear about all the things I saw. He told me some crazy things he had witnessed, like an excited little kid. I picked up a couple of the leaflets, and on one side, it showed a soldier lying in his arms, and in the next panel, he was happily embracing his family. On the other side of the leaflet, it showed a soldier raising his weapon, and in the next panel, the same soldier had comical letter X's over his eyes. In cartoon parlance, this means dead. It was a straightforward story: you put down your guns, you live; if you don't, you die. That explained why there was military equipment scattered all around. They read the leaflet and disarmed themselves. They even left their boots in some cases, which baffled me, but it clearly stated *that all military equipment was required*. Some soldiers took that to mean boots.

Finally, we formed up, and Sgt. Steward said we would break off into pairs, form a line twenty meters apart, and walk through the village. I was paired up with Martinez, and we were towards the end of the line, which meant we were on the outer edge of the village, twenty meters apart. I was told we would have buildings between us and the next pair of soldiers.

Women and children were still in the village; some wouldn't leave. Their experience shook them, and they were unfortunate to see. I saw no soldiers. There could be nothing in this village, and we began moving slowly. We had never trained for urban warfare, so clearing rooms and buildings was a new experience. We entered the first building, which was empty, with nothing inside. We touched nothing, and we didn't open cupboards.

Booby traps could be anywhere; I learned this from the war films I grew up watching. Then, instead of waiting for Martinez, I walked out the door while he was still looking through a bag of mail he had found on the floor. As I exited the building, my eyes caught a glimpse of something moving in the corner.

I spun around to see a young boy, no more than fourteen, standing just a few feet away. His face was devoid of ill intent, a mere reflection of innocence. He had emerged from the side of the building, and I noticed he was empty-handed. But my heart sank as I watched him lunge for something leaning against his leg. Instantly, he pulled out an AK-47, and my world seemed to come to a standstill.

As he aimed the weapon at me, I instinctively knew I was about to die. A rush of sensations flooded my body, and my muscles began to relax. However, my M16 was still slung over my shoulder, and I knew I couldn't draw it fast enough.

But then, just as the boy was about to pull the trigger, there was a distinct click. The weapon had misfired, an occurrence that was rare with an AK-47. The boy tried to reload, but I was already moving.

In a blur of motion, I covered the fifteen feet between us in mere seconds, my M16 finally in my hands. I grabbed the barrel of his gun with my left hand, pushing it away from me while snatching the weapon out of his grip with my right.

The boy didn't have a chance to react before I hit him with the butt of my rifle. I swung my upper body back as far as possible and snapped forward, striking him across the jaw. The impact was bone-shattering, and he crumpled like a ragdoll, sliding three feet across the desert floor before stopping.

Everything around me seemed muffled momentarily, and I could only hear my breathing. But then, the voices of my comrades cut through the silence, and I realized they had heard Martinez shouting. I had acted without thinking, but I knew that I had saved myself and the lives of my fellow soldiers.

The memory of that moment has stayed with me to this day.

Several soldiers ran up from my left, locking and loading their rifles. "Shoot that fucker in the head!" they shouted.

I was still in a daze, but I did not want to kill some kid, even though he had tried to murder me—very uncool. I guess revenge did not come naturally. I still don't know who yelled that, but I was in a state of shock. I was so mad at myself for having my weapon slung over my shoulder.

Martinez ran around the side of the building, where the kid must have been sitting when we arrived, because he hadn't been there when we kicked in the door and went in. No one was on the other side, and Martinez said we were clear. Another soldier ran up, locked, loaded, and put the gun to the kid's head as he lay on the ground.

Martinez yelled at him to sling his weapon. The soldier yelled that we should end this Iraqi piece of shit. Martinez wasn't hearing it; it was a kid.

A medic came running up and told the soldier to stand down, which he did. Then, he began to examine the kid. I grabbed the AK that I had disarmed from the kid. I walked over to the side of the building, pulled the charging handle back, and released it. A round entered the chamber, and I discharged it into the air.

It was a jam. It shouldn't have jammed, but it did. I began to relax. My breathing was still heavy. The voices started to become more apparent. My focus returned.

Martinez set his hand on my shoulder. "Are you okay?"

"I'm fine. Where the fuck did he come from? How old was he?"

"He must have been on the other side of the house. Don't know, didn't clock him."

"I was dead," I responded.

"You're not dead, Joe. Take a minute, and let's go through the rest of these buildings. Do you wanna head back to the track?"

"No, let's get this done," I replied.

The kid was dragged away. I have no idea where, and I don't care. All the different voices coming at me were overwhelming, but things were settling now.

Sgt. Steward came to check on me. I was shaken but okay. I drank some water. My adrenaline was spiking. No one saw me with my weapon slung, only the kid. I never got yelled at for that mistake. Instead, I carried that anger at myself for years. As a result, I never felt luckier, and I should know luck; according to my email, I've won the Nigerian lottery over twenty times.

Sgt. Steward got us all formed up, and we started searching for the rest of the buildings. I was still with Martinez. We had a few buildings on our right to clear, and some were a few hundred yards away to our left. I still didn't feel right. I was coming down.

The hyperfocus was still there, as if seeing the world in high definition. I honestly think all my senses were heightened.

I still never felt any fear of death, just acceptance. It was a good run. I was nineteen and had so much left to do, but I felt it was okay and everything would be all right. We continued to inspect the buildings but found nothing out of the ordinary. It took about twenty minutes to clear the buildings on our right. We looked through each room, but there were no people, nothing extraordinary, and no soldiers.

Finally, we got to the building to the left. We decided to split up and kick in both doors simultaneously. Like before, I kicked in my door and rushed in, but there were no people. I looked around, and back against the wall were two large crates with Cyrillic writing on the side. The arrows are pointing up. I started to feel like myself again. I searched the whole building; it was just one giant room, and my attention turned back to the crates. The floor was dirt. I closed the door and kicked it in so nobody could come up behind me without them hearing me.

The crates were huge and nailed shut. My entrenching tool was on me, so I unfolded it into a mini-shovel. I used the shovel to pry open the top of the crate. I was ignoring my advice. Haphazardly opening the crate, overcome with excitement. Hey dumbass, this could be wired. My excitement overrode my instincts to wait for Martinez, but I didn't; I ripped the top off. I could hear Martinez outside talking to some other soldiers about something. I was not paying attention; I focused solely on what was before me. After the top is off, I am met with a brown gelatinous material. I examined the edges of the crate, looking for any wires, and then I looked behind it. I see nothing. I explore the bottom corners and brush away the dirt to see if anything is buried underneath. I find nothing. It isn't a trap. I examine the brown goo, wipe my finger across the top, and sniff. It's grease. What the fuck are they packing in grease? Truck parts for some vehicles are likely scattered throughout the village.

I unzipped my MOPP suit, hung it around my waist, rolled up my T-shirt sleeves, and reached into the grease. I felt around and grabbed onto a pipe. It felt like a pipe, but then I thought about the top part of the pipe and realized it was a barrel. So, I held it and struggled to pull it out of the grease. I finally got it out, and it was a brand-spanking new AK-47.

I reached in again and pulled out a second one. They were not loaded; they were just weapons with no magazines. Luckily for me, there were tons of AK magazines lying around outside.

Martinez came in and started yelling at me. "Why did you open that crate? It would have been best if you had come to pick me up. No less than twenty minutes ago, didn't you tell me not to open or grab anything? Take your advice, kid."

He was right, but I didn't care. My brain was not even fully formed at that point. I was already wiping the grease off the AKs, with plans to grab some magazines and shoot them off into the desert. Martinez cracked open the second crate and found more of the same—containers filled to the top with grease and packed with AKs.

I grabbed my two new rifles, went outside, and showed them off to everyone. This caused a massive frenzy to get into that building, but the higher-ups stopped everyone from entering. I walked through the village and found some fully loaded magazines lying in a pile of equipment. I hit the magazine against my helmet to remove any dirt. I loaded a magazine into each rifle.

I had my M16 slung on my shoulder and a shiny new AK in each hand. I headed out to the edge of the village, my unit behind me. I fired both machine guns into the empty expanse, like Scarface. I couldn't have been happier at that moment.

I forgot where I was. I heard some of my friends coming up, thinking I was about to get yelled at, but no, they were carrying more magazines for me so we could all take turns with my new rifles. That was peak living for a nineteen-year-old male who grew up watching 80s action films and living through the Reagan pro-military years.

I felt like Rambo or Rambo adjacent, at least. I felt so free to shoot guns, and nobody stopped me. Did I scream while shooting in the air with abandon? You bet your ass I did, and it was as satisfying as it appears in the movies. I was used to the range mentality; everything was monitored, but I could fire off here, and everyone was genuinely excited.

I hid my AKs in the toolbox on the track. I shot them off simultaneously, so I achieved peak awesomeness. After almost dying twenty minutes earlier, I had moved on from that horrifying life experience to feeling genuinely alive, firing off those weapons.

The peaks and valleys of war all in twenty minutes. We loaded back on the track and got to the edge of the village. It was late afternoon, and I saw some of my other squad members for the first time.

Finally, we had time to wash up and relax before sweeping back through the village. We assembled and ate chow, telling stories. *What did you see? Did you see this? Did you see that?*

By that point, everyone seemed to be run down. We were each handed new MOPP suits to change into; we'd been wearing our suits for a month since January 16. We had no access to showers either; you took a whore's bath—wipe down the vital areas.

We pulled off our suits, grabbed a water jug, and cleaned ourselves. The cold water hitting your skin was an indescribable joy. You didn't know how much you missed feeling clean. I was covered in charcoal, which the MOPP suit used to prevent chemicals from coming into contact with my skin. I washed my face and body—the new, fresh suit- and cleaned underneath the suit.

I felt great and refreshed, and we gathered our gear, split up, and walked back through in pairs. Not all of us, just a small group, including me. We moved to the outer edge on the opposite side of where I was when I encountered that kid. We had to evacuate everyone who was not planning to leave the village.

We didn't care where they went, but they could not stay there, especially after finding those stashes of AKs. I told some women to move, and Martinez saw a guy sleeping on the upper ledge of an incomplete house. He told me to get that guy to move. I searched fifty yards toward him, scanning for people and mine, but found neither. The guy was lying on the ledge, about four feet off the ground.

I'm six feet, and I had to raise my rifle above my head to poke him. "You can't sleep here, and you have to move."

He didn't move. I heard a noise coming from him, but there was no movement. Gurgling? This guy was sleeping on jagged rocks, curled up in a ball. He looked so comfortable; I wished I could sleep that soundly outdoors.

I poked him again. "Hey, let's go, move! I will pull you off that ledge."

I was now wondering how he even got up on that ledge. I was getting more agitated.

"I'm going to pull you off this ledge, and it will hurt," I warned him. "This is on you!"

I knew he didn't speak English. I was shouting this basically into the void. I slung my weapon and grabbed his shirt beside his shoulder with my left hand and his pants near the ankle with my right. I pulled, and—

As he turned over towards me, I saw that nothing was holding his insides in, no skin or rib cage. It was an open wound; his entire chest cavity was exposed. He had no face. It was a skull. All his insides spilled onto me. I'm talking lungs, heart, stomach, and intestines, all outside where they certainly don't belong, and they were all over me. Some of his organs got caught in his exposed, broken rib cage. They hit my helmet, my face. The rest covered my new, clean MOPP suit. Blood soaked into my new, clean suit, staining my helmet where something wet and solid fell out of his stomach cavity and hit me on the head. *Is that a human liver on my arm?*

He took a six-foot fall from his sleeping perch, and a giant splash of guts and viscera exploded in an upward motion all over me when his body hit the ground. He was severely burned, too. How the fuck did he get up there in this condition?

I was covered head to toe in blood, like Carrie in the school gym. I was furious.

I must have yelled, too, since Martinez ran over to see what was happening.

"I just got clean!" I yelled.

I kicked the body out of anger a few times, not my finest moment. When Martinez saw the disfigured body, he even recoiled in horror and had to collect his thoughts. I couldn't believe this shit. Martinez told me to go back and ask for another MOPP suit and clean up again, and jokingly told the man he had to leave the area. I was sure he was trying to make me laugh. He was such a good guy.

I returned to where my truck was, and everyone saw me coming, covered in blood, and asked how many people I had slaughtered; I was only supposed to tell them to leave, not kill them. I got a new MOPP suit, cleaned my face of all the blood, and cleaned my helmet as much as possible.

I was told it would be a while before I could get a new helmet cover. After a few days of drying, it eventually became a big brown stain on my helmet.

I couldn't get clean enough. I kept thinking there were still blood spots on my skin. I was the Lady Macbeth of Iraq. I could not get clean. After cleaning myself of blood and intestines, we started corralling refugees and soldiers who had given themselves up. They were now Prisoners of War (POWs). We searched for weapons and lined them up. A couple spoke English, including an Iraqi colonel who attended school in America. We fed them; they were starving. We were told they had water rations and had survived on a small rice bowl for the last month. They also had no idea where they were, and we had driven two days before encountering our first bit of resistance. The soldiers thought they were on the border.

I'm sure their colonel knew better. I listened in while they interrogated him; he was very forthcoming. He was looking forward to eating a nice meal like his troops. He kept eyeing them, sitting and eating a few yards away. There were refugees everywhere, women and children, covered head to toe in dirt and grime, hungry and distressed. Witnessing it was tough, and the children's crying filled the air.

Mothers tried to comfort them, and we fed as many as possible, handing them MREs. I gave a group a box of them out of our track vehicle. They needed it more than we did. These people didn't deserve this; they didn't ask for it, which left an indelible mark within me that I still carry around.

It was getting dark, and we planned to meet with the rest of our unit. I would finally see some of my other squad members who had separated from us when we crossed into Iraq. We set up our tracks in a perimeter.

We couldn't be more than five hundred yards from a long line of howitzers, which began shelling where we were going the following day. There seemed to be a lull in fighting at that point. I don't remember any gunfire or explosions.

We could see oil derricks spewing flames in the distance, and they appeared to be twenty miles away. Memory is fickle. My buddy took some great pictures of them burning.

We were told to get some sleep, and the guard schedule was set up, like always. Sure, I'd get some sleep. It just did not seem possible. I felt like I was losing my mind at this point. Sgt. Steward told us we were changing course and heading toward Baghdad the following day.

We were not far from Basra, a major city in Iraq located near the border with Kuwait. This meant we were heading north the following day. At that point, I feared that war in an open desert where you could see your enemy miles away differed significantly from urban combat. Buildings, open windows, snipers, and countless civilians are a different bag of fucked up. I was unsure of the plan, but I could feel the three days of very little sleep.

I remember hearing things like gunfire that was not there. Sleep was the only remedy. I made a little hooch on the side of the track with my poncho and put my cot beneath it, like always. The howitzers did not stop firing for the next six hours; the earth shook with every explosion. It was impossible to sleep, but I'm sure I dozed off at some point. I did guard duty and was woken up because we were on the move again; this time, our new heading was Baghdad, possible urban combat, which we were not trained for. This is not going to end well.

Chapter 11

I woke early on February 28, 1991. The sun was coming up. The howitzers were silent for the first time in hours, leaving only an eerie silence. You become so accustomed to the noise and repetition that the silence is deafening when it finally stops.

It was another grey morning. We would get hot chow before we left, which was fantastic. I needed coffee. I grabbed my gear and my rifle and headed to chow. As I was walking toward the truck, a soldier approached me, an officer.

As he got closer, he smiled from ear to ear. "The war is over. You must tell your squad leader to turn the track around. We're heading home."

I was in a daze, without coffee or sleep, and hadn't jerked off in days. I was in no mood. "What? What do you mean the war is over?"

"I mean exactly what I mean," he said with a deadpan look. His voice was steady, with no excitement, probably just as tired as I was. "It's over. We're going home. We need to turn these tracks around. So, send your squad leader over to the TOC."

Could we have done that the entire time? We could have just turned around and gone home. I would have preferred that stance. I excitedly ran back and told everyone that the war was over and we were going home. No one believed me. I told Sgt. We needed to head over to the Tactical Operations Center (TOC). He returned and verified that what I said was true, and we were all going home.

I remember every face covered with black soot or dirt, but you could see everyone's smiles. Music started blasting. I heard "Brothers in Arms" by Dire Straits from one of the Humvees and sent it over the radio. I sat there for a few minutes, listening to the song and looking out the front. It was a grey-blue sky. I could smell the artillery shells. They smelled good.

I remember the strong smell of coffee. It was just so peaceful now. I could hear cheering near the chow truck. No gunfire or howitzers were firing off, just quiet. The silence was deafening. I thought it might be another Vietnam—do a year in the desert—but hearing this news after only four days was like having a weight lifted off me.

The infantry was going to start heading southeast, and we would follow. I was too tired to cry out with happiness. I survived it. I didn't think I would, but there we were, and I was going home.

But we were not out of the woods yet. We were deep in enemy territory and still had to return to Saudi Arabia and finally home. A lot could happen during this time, and we had to be cautious.

The engineer's work was never done. The infantry was kicking back in the shade. We were tasked to drive, blow up mines, put razor wire around unsafe areas, and dig ditches. Our other mission was gathering enemy equipment. This was left in small piles all over the desert. We walked around with a Humvee or truck in front of us, picking up bullets, grenades, boots, ammunition clips, and everything else dropped on the ground. I was walking around with a group of five or so guys. Caldwell was there, running his mouth.

"My daddy this... My daddy that..."

We came across a box of Russian grenades in the middle of nowhere. They needed to be used. Instead of throwing it in the back of the Humvee, we decided to fuse the grenades ourselves and randomly throw them in the desert, you know, for fun. What nobody knew about these particular Russian grenades is that when you pull the pin, it makes a popping sound, like an old cap pistol. This startled the first guy who drew the pin and, out of fear, made a half-assed throw right over the other side of the dune instead of throwing it far. It was cooking off eight feet away from us, and the box had six other grenades.

We all dove to the ground. I covered my head and plugged my ears, and it exploded. I have never wondered why women live longer than men—this is a prime example.

I felt the impact throughout my body; it shook my bones. There was a ringing in my ear.

We all got up and started chasing the guy who threw it. Shaking off the almost fatal accident we just experienced took a while. Once we recovered, we fused the remaining grenades, pulled the pins, and threw them as far as we could in the open desert. Two or three grenades went off at the same time. No one was supervising us; it was so much fun, other than almost dying from that first grenade.

We continued to move forward, picking up truckloads of military equipment. We dumped all the equipment in the giant hole that ACE Earthmovers had created, which was filled to the top with guns, grenades, and ammunition. We slid Bangalore torpedoes through the equipment. We were testing our timer cord, which gave us five minutes to get to a safe distance. We placed C4 everywhere, attaching blasting caps to the end of the detonation cord.

We laid dynamite; nothing was left to chance, and the equipment needed to be in tiny pieces. We tied a girth hitch knot of the timer cord around the det cord and pulled the fuse. We hopped in our Humvees and drove back to where the rest of my unit was waiting. We cut the timer cord after a test burn for about five minutes. The anticipation was killing me. We all hoped it didn't misfire.

It didn't. We were a mile away from the blast.

At that distance, the shock wave took ten seconds to hit us. You saw the explosion and then, seconds later, heard the boom. There were so many explosives in that hole that it made a mushroom cloud.

CHAPTER 12

The ceasefire was signed. Another day passed, and we were probably all too relaxed. We had fallen into a routine. We still had to work. They wanted us out looking for these mines that were maiming and killing people because they looked like metal origami spiders, cute and deadly. Kids would pick them up, thinking they were toys.

We spent a few hours setting charges next to these mines. While doing this, we encountered a dirt berm about ten feet high. It was about 2 p.m. There were fewer people today; it was just me and some lieutenant in a Humvee. I could see other engineers to my left and right. I walked over to the berm. Track marks and oil patches were visible on the other side, along with the tire marks of trucks. This berm was miles long.

We had found an Iraqi rally point. I climbed to the top of the berm and gazed down at the lieutenant, who pointed behind me. He threw me some binoculars, and tanks and trucks were coming straight for us. It was a Lawrence-of-Arabia-esque scene. The horizon was filled with dust clouds, with hundreds of trucks and tanks barrelling towards us.

I learned later that those elements of the Republican Guard's Hammurabi Division engaged my unit, the 24th Infantry, and the 1st

Brigade just north of Highway 8. They knew there was a ceasefire; they didn't care.

Finally, we heard chatter on the radio to pull back. I jumped in the Humvee, and we sped back towards our unit. I saw M1 Abrams and Bradley Fighting Vehicles heading in opposite directions toward the tanks.

We passed them, drove up this giant hill, and parked at the top. Big rocks lined the top, and the lieutenant said we could watch the battle from behind the rocks. We each had a pair of binoculars. He turned the radio in the Humvee way up so we could hear the play-by-play.

I got to watch this incredible tank battle. Iraqi shells bounced off our tanks and exploded on the ground next to the tanks. Watching our old ammunition not even dent the tank armor was comical. Apache helicopters joined and blew trucks and Iraqi tanks up with missiles. A10 Vulcan planes straffing targets in the valley in front of us. Troops scrambling out of burning tanks and trucks, getting picked off by helicopter fire. We did not drive that far away. The hill we were on overlooked the entire valley of the battle. Iraqi tanks kept getting closer and closer to us until a couple were below the hill we were perched on. I knew we had a rocket launcher in the Humvee. I asked the lieutenant if I could launch a rocket at one of the Iraqi tanks below us.

"Sure," the lieutenant said. "Go get the rocket launcher. It's in the back."

I ran to the Humvee, opened the case, and pulled out the launcher. I extended it, reading the instructions on the side to ensure

I did everything right. I had to pull a pin before it fired. No one could be behind me; there was a sixty-foot backblast.

I kept low behind the rocks. An Iraqi soldier jumped out of the hatch and had a machine gun. We were no more than two hundred yards above the tank. They had no idea we were there, and I pulled the pin on the rocket launcher.

I stood up, and now I could see the tank. He still didn't see me. They began to move, picking up speed and moving away from me. I felt like I was missing my chance. I barely aimed and pulled the trigger—the rocket zigzagged to the tank.

I missed my shot, and there was a small explosion of dirt in the desert. I couldn't be more disappointed in myself. I sat down, deflated. My one shot, and I blew it. This was probably my peak coolness in life, and I blew it.

As I stood there, I could hear the tankers taunting the Iraqis over the radio, knowing they had the upper hand. This was not a fair fight, as the Iraqis were grossly under-armed. After the battle subsided, we drove down to the site to survey the aftermath.

The scene was devastating, with tanks and trucks smoldering in the aftermath of the war. I held my rifle tightly, prepared for any surprises that might come our way. One of the tanks was smoking, but it didn't appear heavily damaged from where I stood.

As I approached the tank, I noticed a couple of holes in the armor, but they weren't significant enough to have caused damage inside. I could hear faint movement from within the tank, and as I climbed up to take a closer look, the smoke billowing out of the hatch cleared. I was horrified by what I saw.

The entire interior of the tank was covered in gore. I couldn't believe my eyes as I counted ears and tried to figure out what could have caused such damage. Part of a human face, with one eyeball, was on the floor. It took a second to register what I was looking at. I stared at it; it stared back at me.

Are there men inside? We listened for movement. Several holes in the tank shouldn't be there. The lieutenant suggested dropping a grenade inside the tank, but upon closer inspection, we realized it wouldn't be necessary. We suspected that a sabot round had caused the damage, which created a vacuum in the tank cockpit, sucking everything out of the exit hole.

The unfairness of the whole situation hit me hard, and I couldn't help but feel sorry for the Iraqis who never stood a chance against the world's most powerful military.

The Iraqis had lost over a hundred and eighty-five armored vehicles and four hundred trucks in just a few hours. As we searched the tank for any intel, I had to crawl inside and navigate through pools of blood, trying not to step on any organs or slip. Blood dripped from the ceiling, and I had to wear my poncho to avoid getting drenched and staining my MOPP suit with viscera.

I wish I had never opened that hatch, but when you are commanded to look inside, you do it, not knowing that what you are about to see will haunt you for the rest of your life. It was an unreal scene; it was overkill, which could have been a metaphor for what we were doing to the Iraqi Army. It was a puppy fighting a bear.

CHAPTER 13

The war lasted a hundred hours, and U.S. and allied ground forces in Iraq and Kuwait defeated a battle-hardened and dangerous enemy. U.S. and allied forces destroyed over three thousand tanks, fourteen hundred armored personnel carriers, twenty-two hundred artillery pieces, and countless other vehicles during air and ground operations. This was achieved at a cost to the United States of ninety-six soldiers killed in action. I only saw our medical teams working on wounded Iraqi soldiers; I only witnessed one US injury.

Kuwait was now liberated—you're welcome, Kuwait! U.S. forces turned to humanitarian missions. They sorted out refugees, assisted the Kuwaitis in reoccupying their cities, and helped them begin rebuilding. U.S. Army Civil Affairs and Corps of Engineers units set up food, water, fuel distribution points, and medical clinics.

I got an early wake-up call from Sgt. Steward in late February, maybe early March. He said he and I were going on a special mission. The military was not going to ship all the ammunition and missiles home. Our job for the next few days was to use all the ammunition and blow up all ordnances not coming home. He tells me we will destroy $22 million worth of Stinger missiles. Here I was, this e-nothing, and he put me in charge of blowing up these missiles.

He said I could do anything I wanted. I started modeling the C4 into gophers, as in the movie *Caddyshack*. They had lowered all the missiles into the hole before we arrived; all I had to do was set the explosives. I used Bangalore Torpedoes, and I put my C4 gophers up everywhere.

I used my knife to dig a hole in the C4 and insert the blasting cap attached to the det cord. I set my timer to fuse for five minutes and attached the timer cord to the det cord. Sgt. Steward watched everything I was doing; if I needed correcting, he corrected me.

It took us about thirty minutes to set everything up. Steward loaded up in the Humvee. I pulled the fuse, jumped back, and started driving off. We were not too far away when I saw a wild pack of dogs running toward the missiles. I pointed this out to Sgt. Steward. The wild dogs were sniffing and searching for food we may have left. They only had less than a minute to live.

"We need to go back and stop them!" I said, with tears beginning to form.

"Joe, we can't. They probably think we left food over there."

"We have to stop it!" I insisted.

"Joe, we can't save everyone. They're just dogs."

"Dogs are better than people!"

That one hurt the most because I saw all the death and destruction. Animals are innocent. They need protection. That just broke me even more than I already was.

Again, the stinger missiles exploded—no mushroom cloud this time, but a monster explosion just the same—those poor animals.

We got back to our unit. The death of those dogs weighed heavily on me, but I saw that our unit set up a range with the husks of Iraqi tanks downrange that we could fire all our ammunition at.

My nineteen-year-old brain shifted on a dime. I now wanted to shoot all my rounds. I saw rockets being launched. 50 cal's lighting up the tanks. I grabbed a can of ammo and started loading clips.

We spent the next couple of days shooting everything we had downrange. Shooting and firing rocket launchers all day was so much fun. It was everything I thought the Army would be. I honestly was being all that I could be.

Sgt. Slevens noticed my AKs. I'd carry them around with me, both slung on my back. He told me I had to turn them into the company armorer.

"Why?" I asked.

"You think you're going to take them home with you? Not in this lifetime, you're not. They are not yours." He says in his most dickish voice.

"They *are* mine. I found them. I believe the law says, 'Finder's keepers.' I believe it is federal. You have another thing coming if you think you're taking them." I reply in my dickish voice.

"No, I'm taking them and giving them to the armorer. This is an order. In case you haven't noticed, I'm your superior." His voice is getting more agitated.

"You are in no way superior to me. I don't rob my troops. You use your subordinates as an ATM; it's bullshit. So fuck you, bust me if you don't like the way I'm speaking to you. I'm E-nothing! What, are you going to bust me down to civilian? Please send me home. Fuck these bullshit rules." I replied, and my voice was agitated.

He stared at me. I hear Sgt. Steward came up behind me and told me to give them up. I reluctantly did. I respect Steward, but this was the wrong move; he should have backed me up. I say this because Sleven did precisely what I thought he would do: he took my AKs apart and mailed them home piece by piece. Mailing things home was so simple; there was no need for postage. In wartime, all we had to do to send a letter was write *Desert Storm* and your signature where the stamp would usually be. You had to write your unit on the other side. The envelope had no weight limit, allowing a large piece to be sealed and mailed without issue, such as a rifle bolt. A few days later, I asked my friend, the armorer, if Slevens had turned in my weapons. You already know the answer. I had no recourse. I'm still pissed about this thirty years later, fuck that guy.

Chapter 14

Our entire unit was reunited for the first time. All the squads and support units, including Sgt. Ramirez, who ran the supply unit and was his usual self.

"Fuck you, faggot motherfucker," was what I was greeted with when I requested a new cover. "I'm too busy to fish one out." Ramirez barked.

I just stood there and stared at him.

"Fine, faggot, I'll get you one."

He wove a tapestry of profanities. He hit every marginalized group. He was an equal-opportunity offender. It was admirable. He even made fun of himself. In the nineties, this word was used so freely, almost like punctuation at the end of a sentence for some people.

I was finally able to get a new helmet cover. The bloodstains on it had now darkened to a deep brown from exposure to the sun.

He smelled the stain. "What is this faggot shit, motherfucker?"

"A major organ fell out of a guy's stomach or chest and hit my helmet. I'm not a doctor, but it was horrific- probably a kidney stain. It is blood. That's all you need to know."

"I will fuck you, motherfucker," he always threatened. But, of course, *motherfucker* was his favorite word. It flowed off his tongue with aplomb. You had to admire how he could weave it into any sentence.

But, of course, he also just said what was on his mind. That was just who he was, no governor in his mouth. He didn't care if you were insulted, and he was coasting until retirement in a few years.

So, as we began our drive back through Iraq, the stench of death was overwhelming. The scorching sun had reduced once-mighty tanks and trucks to smoldering wrecks. Once filled with hopes and dreams, bodies lay decomposing inside them, their skin melting off their skulls. Some had burned up entirely, their corpse hands still gripping the steering wheel. ACE Earthmovers were already at work, digging shallow graves and unceremoniously pushing the vehicles and their occupants into them.

I couldn't help but think of the families of those lost soldiers who would never know where their loved ones lay buried in the unforgiving desert sands. I couldn't help but see these people for who they were, not just faceless enemies but human beings with their own stories, fears, and dreams. I knew many were likely there against their will, forced to fight by their government. The thought of it made me feel even sadder. It could have easily been me in their shoes, lying in the desert with no ceremony or final goodbye for my loved ones.

Basic training might have tried to purge my humanity, but it hadn't succeeded. I wondered how many had been robbed of their futures and forced to fight a war they didn't believe in. The senseless loss weighed heavily on me as we drove on, leaving the shallow graves and the ghosts of the dead behind.

Yet, as we made our way home, I felt the burden of hyper-vigilance slowly dissipating. The constant threat of danger, always at the forefront of my mind, was finally lifting. I fell asleep on the back of the track, exhausted beyond measure. I slept for almost fifteen hours, a deep, dreamless slumber from which the rest of my comrades couldn't even wake me.

I fought with one of my teammates in my sleep, who tried to wake me violently. Eventually, they just let me sleep, and for once, I didn't have to pull guard duty. Understanding the toll that war had taken on me, Sgt. Steward allowed me to rest.

We formed our perimeter, as always. They were the most U.S. vehicles I'd seen since the war ended. Our fleet of mechanized vehicles was parked in the port. I got to sit in the driver's seat of the M1. The tank commander told us that if something chipped the tank's armor, they had to bag that piece up and send it up the chain of command. The composite armor was so secret that they couldn't have it fall into enemy hands.

I watched as more rounds bounced off the armor and exploded next to the tank, causing no damage to the vehicle or crew. These were right out of the factory and modified in the country. The machine gun was laser-guided and controlled by a computer, which was already on the next target while illuminating the current one.

We waited as a unit for the trucks to arrive and take our track vehicles back down to Dhahran. Instead of us driving ourselves back, the US Army hired a fleet of trucks with a huge flatbed trailer to load the track vehicles on, and the Egyptians would drive us back to Saudi Arabia. As we waited, we were back in hurry-up-and-wait mode, the default setting for any military unit. We were cleaning weapons. All the vehicles were receiving some form of maintenance. The track drivers had to pull out the giant air filters and hit them to knock all the sand out. They usually did it once a day. We all chipped in. Once all the busy work was done, I dove into a John Grisham novel; we had the time.

The trucks arrived, driven by these teams of Egyptian men. To prepare, we began unloading all our tools and explosives onto the back of the deuce and a half trucks.

We needed to lighten the load of the tracks so they could be driven onto the flatbed trailer, two tracks per trailer. We also discovered that two men would be in the back of each track. Bunk would be my trackmate. This is the most unsafe thing I've ever heard.

My track is the second one on the trailer. The flatbed shocks flatten out and make weird creaking noises when the multi-ton truck rolls onto the back. The sounds coming off these trailers couldn't be more unsettling. Did we empty enough equipment? I was sure we were going to die in transit back to Saudi Arabia after surviving the war. This seemed the unsafest way to transport our equipment and troops back home. The tracks looked like they would fall off the truck at any second.

Bunk and I load up in the back of the truck; this is the cleanest the truck has been the entire time I've been in the country. We swept out the sand and had two empty wooden boxes to lie on, which were uncomfortable. I rolled out my Army bedroll. I'm sure the quarter-inch of foam will make this wooden box tolerable to lie on. I was wrong; lying on this thing felt like a Guantanamo Bay stress position. Even lying on top of my sleeping bag didn't help. Bunk is immediately asleep on one of the wooden storage boxes. We opened the top hatch to let some air in so I could stick my head out and see what was happening. My adrenaline was pumping, so I couldn't sleep, and my fear of these trailers tipping over was never far from my thoughts. The Egyptians jumped in the truck's cab, and we formed our convoy and were off. As we bounced along the dusty roads, I couldn't help but feel like we were careening toward disaster.

The Egyptian drivers took off at breakneck speed, their recklessness making my heart race with fear. I'm in a hurry to get home but not to die. The trailers we rode in seemed impossibly top-heavy, and I braced myself for the moment when they would tip over, sending us all flying out into the unforgiving desert.

The first turn onto this long dirt road was sharp, and the trailer in front of us was tipping to the left. It looked like it was about to tip over at 30 miles per hour. The driver straightened out, and the trailer straightened back up.

My heart started pounding again as we swayed precariously on the backs of those trucks, five feet off the ground, with nothing but the wind to hold us steady. Reaching our destination took us close to 24 hours of non-stop driving. This probably would have taken 15 hours had we not stopped for the drivers to pray every few hours and eat massive amounts of bread and dates, which seemed to be their primary diet. The drive was dusty. I put on my headphones and tried to push those tipping-over thoughts to the back of my mind. The entire time, we are jerked around the back of the track. It was a miserable experience. It was scarier than almost anything else I experienced in war.

As we pulled into the American-controlled part of Dhahran, I felt a sense of relief wash over me. The fenced-off section of the city was a welcome respite from the dangers of the open road. Finally, we arrived at Khobar Towers, a complex of luxurious condominiums purchased by the U.S. government to house returning troops. Seeing such luxury in the middle of a war zone was a strange sight.

However, despite the grandeur of the surroundings, the food consistently fell short of expectations. Pizza Hut was a disappointing representation of what we were accustomed to back home, and the other restaurants needed improvement. I found myself eating camel fried rice and Italian food most nights, anything to avoid the weird-tasting cheese and fake pepperonis of the "American" pizza. The ice cream tasted like ice cream; hard to fuck that up.

Nevertheless, I made love to that cone. Finally, we were given our assignments, and I ran to the building where our unit was staying. After almost two months without one, the thought of a shower had me practically drooling. I slung my weapon and duffel bag over my shoulder and sprinted to the first bathroom, desperate to wash away the grime and charcoal that clung to my skin.

The Khobar Tower condos were stunning, with gold fixtures and marble floors gleaming in the light. They were way better than the barracks where I would live in Georgia—no marble, just cement blocks with a cement-like floor. Our barracks were painted in segregation white, like an insane asylum.

I was faster than everyone else and locked myself in the bathroom. I could hear my fellow soldiers pounding on the door, urging me to hurry up, but I didn't care. This was my moment of respite, and I intended to savor every second of it. If I had a candle, I would have lit it and taken a bath, but that was not to be—just a long shower. I was showing indoors, no less.

The bliss of that hot water touching your body after two months is challenging to describe. I never thought the sight of lathering soap would almost bring me to tears. I looked at my skeletal frame in the mirror, covered in charcoal dust. I had lost much of the muscular definition I had gained from basic training. No matter how little I weigh, I've never seen an ab on my body. I was forty percent charcoal dust and probably didn't weigh over one hundred fifty pounds.

I smelled like a decaying corpse. My shower lasted twenty minutes, but it would have gone on for an hour if I had been in charge. You would think I was reenacting a shampoo commercial by how many times I ran my fingers through my hair.

As I walked down the hallway of our barracks, the sounds of water hitting tile echoed off the walls. The rest of my unit was busy using the showers and other bathrooms on the floor.

My friend Juan called to ask me to check his back for leftover charcoal. I was obliged, but upon inspection, I realized that what I thought was charcoal was a birthmark. Poor Juan relentlessly showered, but the stubborn mark refused to budge.

I couldn't escape the charcoal. It clung to every crevice of my body, even ones I didn't know existed. It was a constant reminder of the dangerous situation we had just escaped from.

When our rifles were taken from us, the reality of our situation set in. The first day at Khobar Towers was uneventful, but it allowed me to call my parents and explore the area. I couldn't resist the temptation of the local restaurants and indulged in some much-needed comfort food. But as I stuffed my face, the guilt weighed heavily on my shoulders. I was struggling to maintain my weight, and excessive eating only worsened things.

As I walked around, I ran into Corporal Bard. He loved to threaten rape every time he saw me; he thought it was funny. This was the first thing he ever said to me. He was out of his mind, and I'm glad he was on another squad. Bard was sitting down and was throwing MREs at kids on the other side of the fence next to Khobar Towers. He was hucking them like ninja shurikens as hard as he could at the kids. They were hungry, and they were getting hit in the chest, center mass, with an MRE coming at them at about 25 MPH. He looked like a madman, playing dodgeball with innocent children. I couldn't believe what I was seeing, and I couldn't help but wonder if the stress of our situation had pushed him over the edge.

Bard looked at me and said, "I wish the MREs' packaging were different colors!"

"Why?" I asked.

"So we know which ones are poisoned!" He then proceeded to beam a ten-year-old in the chest with an MRE. He didn't care. This guy was never right in the head. I could never tell whether it was gallows humor or his intrusive thoughts getting the better of him. He had been the same before the war and remained essentially unchanged afterward.

With no strict rules or PT requirements, we were able to enjoy some much-needed downtime. I took advantage of this by catching up on some sleep (eyelid maintenance) and lounging around reading old newspapers and magazines, still nothing newer than two weeks old. Our Captain was pretty cool about this, just letting us recoup. The rest of our days were filled with busy work, the Army way. As the days went on, monotony began to creep in. How many times can I clean my rifle in a week? I'm cleaning my web belt and canteens for the third time this week. They are clean! What are we doing? There was nothing to do. I would get stir-crazy quickly and leave to walk around and straighten my head.

During one of my walks, I stumbled upon the solution to my boredom problem. I was in the Khobar Towers complex when I saw a mail room with a cardboard sign that said, "Ship your duffel bag home today!" This room was barely the size of a closet. In the back of the closet are stacks and stacks of packed duffels about to get shipped home.

I approached the mailroom worker and asked, "What does this mean?

"Are you retarded? It's clear: load your duffel bag with all your equipment and then padlock it up, and we will ship it home for you. Ask someone what you have to keep for the plane home."

I suppose I was assessed on my reading and comprehension skills.

I returned to our temporary barracks and asked the Executive Officer (XO) what we needed to keep. I didn't tell him why I was asking. He would have probably said no, but it's always better to ask for forgiveness later than to ask for permission now.

He said we needed our uniforms and web belt with the canteens. They would be taking photos of us when we got off the plane. Our rifles would be shipped back separately. So, we had to bring our stuff pressed and wear our clean equipment, including our Kevlar helmet. He said nothing else.

I headed back to my bunk and started stuffing my duffel with all my belongings, setting aside the uniforms, underwear, web belt, canteen holders, and now-empty canteens. I kept what I needed. I asked another soldier for the address of our barracks back at Fort Stewart, and he provided it to me. I hefted my heavy duffel, headed to the mailroom, and shipped it off.

Here's the thing: I never asked anyone. I didn't go through the chain of command. I didn't tell Sgt. Steward what I was up to. I saw an opportunity to stop all the busy work that they were heaping on us day in and day out. They were making us clean our equipment repeatedly. *How much cleaner can my canteen get?* I decided to ship my belongings back to the base in Georgia so I wouldn't have to clean my equipment again and would have more free time to explore Saudi Arabia.

The mailroom didn't even question me; they just took my duffel, wrote the address on the side, and threw it on the massive pile behind them. I was the only person in my unit who had done this, and word was spreading. I made a big show of it.

After shipping everything home, I returned to the barracks and ran into Sgt. Steward.

"Joe, where is all your equipment? Was it stolen? I noticed everything was gone, and I've been looking for you. I wanted to make sure you were okay." He asked, his face a puzzled look.

"No, it's gone, shipped home to Fort Stewart," I answered.

"What? What do you mean, shipped? Having your equipment ready for the plane ride home would be best. Who told you that you could do this?" he asked.

"I have everything I need for the plane ride home. I asked the XO what I needed, and he told me, I have it all right here. I also have spare uniforms. The mailroom downstairs has a sign that says they will ship your duffel home." I replied.

"No, you need to ask me. This goes up the chain of command. You don't just mail all your stuff home because you see a sign. We have plans you're not aware of; you need permission. Can you get your things back?

"I don't think so. I saw other people from our unit doing it, so I thought it would be a good idea. The equipment is clean; I've cleaned it three or four times. I don't think it can get any cleaner; this is busy work."

"Everything in the Army is busy work! Grow up!" he exclaimed. He never raised his voice at me, but he seemed pretty pissed. I knew all of this; I just pretended to be ignorant. I also knew there was nothing he could do about it.

He told me to take him to the mailroom, which I did. I could tell he was seething. He was just annoyed that I did things without asking first. As we walked down, I could see him calming down, and he started using his Dad voice towards me. Once we arrived at the mailroom, all his anger dissipated, and he spoke to the attendant and immediately decided to send his duffel, too. We both walked back upstairs, and he asked what he should pack and what he needed for the plane. I suppose he didn't need to ask anyone, as he took his belongings downstairs and shipped them without saying anything to me.

Once the 'duffelgate' died, we all marched to the port and formed up. It was near the end of February or the beginning of March. We were told most of us would fly home in a few days. This news was met with cheers. You could feel everyone smiling from ear to ear. We made it, we survived it, and we went home. We knew it was happening, but hearing it and knowing exactly when we would see our friends and family again was different.

The bad news was that some of us could not take that flight. He began to read the names, and my name was called. We were separated from the rest of the unit. I had expected this since my team got to Saudi Arabia in September 1990. I didn't join the squad until November. I had been at airborne school when the rest of my unit shipped out in September. That still didn't stop it from feeling terrible; all your friends were heading home, and you were stuck behind. The unit was dismissed, but we were kept back for a briefing on our next steps.

The vehicles in the port all needed to be cleaned. It wasn't just a quick hose down, but cleaned to U.S. customs standards. We were working twelve hours on and twelve hours off. It was a small group. They told us the next flight we could get on was April 10. This meant the battalion's fleet of vehicles had to be clean by the ninth of April to get on that flight home. So, I told the officer in charge that we would be on that flight.

The rest of the 3rd Engineer battalion flew back to Fort Stewart the next day, and the nightmare of cleaning these vehicles began. My friends were gone, and I was stuck with many people I did not know, so it was a lonely month. I had to do many things alone. We had to begin working our 12-hour shifts the next day. I had almost no supervision; no one cared where I was. It was great. I spent the other twelve hours sleeping or walking around Dhahran.

Growing up in a border town like San Diego, we were forty minutes from the U.S.-Mexico border. We used to go down to Tijuana. You saw insane poverty, homeless women and children selling oranges and candy. It made you feel for them and how families littered the streets on your way back to the US border.

Walking around Dhahran in Saudi Arabia was next-level poverty I'd never experienced until that point—post-war poverty. Iraq had been at war with Iran for a large chunk of the 80s. This shopping mall was in Dhahran. Hundreds of people begged outside. People were missing limbs, walking around with crudely made crutches, Tiny Tim style, and legless people were wheeling themselves on DIY carts. I saw a man with a large chunk of his skull missing, no eye, and blind in the other eye, walking around with a cane, asking for food. The mall security would see me coming and start hitting and kicking these poor people to make a hole so I could enter the mall. I told the guard not to do that; it made me angry.

I probably should not have gone inside, but I did. The mall I walked into was air-conditioned and opulent, like the Khobar Towers. Marble and gold were everywhere, and Liberace-esqe is the best description.

The guilt I felt eating in the food court after what I witnessed outside is hard to describe. I had money since I had not touched my bank account in six months. I was getting my regular pay and airborne and danger pay. I was in a mechanized airborne unit, getting airborne compensation despite not having my wings from airborne school.

The world was a different place because I walked around Saudi Arabia without a rifle and never felt any danger or unsafe. I wandered through markets. I spoke to people, and they were very kind. I stuck out; I was easy to find. I was taller than most people walking around, making spotting places I wanted to visit much more accessible.

Americans stick out, and I would not walk around those same areas today. I would fear being kidnapped. It is such a different world. I walked around without fear, and everyone was friendly to me. I'm not sure how it is now, but there was minimal American culture to be seen, maybe a movie poster here and there, but no McDonald's. I never found it if it was there. American culture is our number one export, and it was nowhere to be seen. It was like being on another planet, but so was living in Georgia, being from California. You could feel their ancient laws.

At the shopping mall, I left my bag one evening. I walked back to the base and realized that it was missing. My friend said I should go back, but I assumed it would be long gone and decided to walk back into town just the same. I retraced my steps, and there it was, leaning on the leg of the table in the food court. No one even looked at the bag. I opened it up and examined the contents, and everything was there. They didn't fuck around with thieves in that country; you'd lose a hand.

When I returned from the mall and showed up for my shift, they worked with us like dogs. It was a team effort, and it was nice and cool in the port. We were next to the Persian Gulf, and a nice breeze was always coming off the water. It was deep blue, cold water. We would jump in at lunch when the day was at peak hotness. We would wrestle and throw each other in when the mood struck. It was so much fun.

My weapon this time was a high-pressure hose. Our job was to remove all dirt and grime, including grease. You would get intimate with every vehicle, jamming your hose in tiny cracks. You would let the water do the work and hit spots until the water became clear and no more dirt was being jettisoned everywhere. Underneath the vehicle, the mud would gather, and it would be so thick and dense that if you were lying in it and spraying the undercarriage, you would form a seal with the pavement and need people to help lift you out of the mud underneath the vehicle. You would open every hatch, door, and crevice you could find. You would find a spot, and it would be 10 minutes before the water would run clear, and huge chunks of dirt and dried mud would fall to the ground.

The inspectors were not kidding around, using white gloves and running their fingers along all the hidden seams of the trucks and track vehicles. If any smudge of grease or dirt were on a white-gloved finger, they would make you rewash the vehicle and call them back for another inspection. This cycle would repeat, sometimes five times, before the vehicle was passed. You then move on to the next vehicle; there were hundreds of dirt-caked vehicles in the port as far as the eye could see; some days, it seemed impossible that we would get the job done at all, much less in a month. I wanted to go home and get out of this country, so I was determined to clean these vehicles.

One day, I was cleaning a five-ton truck, lying underneath it with only an hour of sleep. The weather was getting hot again, and the cold water splashing on me from under the vehicle felt terrific. Underneath this vehicle was a divot in the ground, so big that I could lie in it. Without realizing it, I fell asleep, and the water drained into the divot, rising higher and higher. The divot slowly filled with water and soon covered most of my body, and my face was slowly submerged.

Someone saw my legs sticking out from the side of the truck and noticed that I was asleep and almost entirely submerged in water. I don't know who, but they saved my life. I was determined to become another statistic, but I was saved again. They grabbed my ankles and dragged me out from underneath the truck. At first, it was funny, but it was utterly unacceptable. I needed to get a proper amount of sleep every night. This is a job, and I need all my faculties to do it. This event changed my whole attitude. No more wandering around; be more professional and get yourself some sleep. Finish this job and go home. I was told to go back to the barracks that day and get some sleep.

My new attitude was noticeable. I became a vehicle washing machine. I would clean and clear five vehicles a day. I would remind everyone every day that April 10 was coming, and we had to be on that flight.

One afternoon, I was in the phone bank waiting for my turn when General Schwarzkopf walked in. The entire phone bank snapped to attention. He put everyone at ease and began to walk down the line, talking and shaking everybody's hands, thanking them for their service. He was a celebrity. I was in the 24th Infantry, and when he saw my patch, he said it had been his first command.

He asked how old I was. I said nineteen, but he told me I looked like I was fifteen. I'd never heard that before. He was undoubtedly a politician. I mentioned that we met at the Faisal range when we did the MICLIC demonstration. It was my vehicle that fired it off. He remembered since it was only a few months back. I was lucky to meet him. He was very much like a politician, but you felt like the most important person when he was speaking to you. It was a brief one-minute meeting, but thrilling for me then.

On April 8, 1991, I headed to the port and started counting the vehicles without a customs sticker. Finally, there was just a handful left. I knew I was going home in two days. I crawled into the back of a covered five-ton truck and cried. That sounds weird, but I didn't know it could still happen. I'm glad I was wrong.

I pulled myself together, hopped out of the truck, and the skeleton crew rallied on those last two days. I remember the looks on everyone's faces when the custom sticker was put on the previous truck. We were going home.

Finally, we got to the airport. I saw the familiar *Tower Air* on the side of the jet. The moment we boarded the plane, I felt relief, and my exhausted body could finally relax. I made it out alive. As the aircraft began to taxi, I could sense the anticipation in the air, and the cheers of fellow soldiers reverberated throughout the cabin. Tears were flowing freely from many, and I could hear the sobs and sniffles of those who were overcome with emotion. But no one made fun of anyone for crying. We had all earned the right to let it out, to release the pent-up emotions we had held onto for so long.

Then came the announcement over the loudspeaker, and a program began to play on the screens in all the cabins. It was a compilation of all we had missed while away from home, featuring the latest music, movies, and news. The comedian Robin Williams was the first thing we saw. He thanked us for our service, followed by his trademark stream-of-consciousness humor. It was a much-needed respite from the horrors we had witnessed.

The video lasted for a couple of hours, and we watched everything from music videos to first-run movies. They even played a couple of episodes of *The Simpsons*, aired its first season that year.

We landed in New York, and as we walked down the empty terminal, we were greeted by a massive table filled with stacks upon stacks of White Castle hamburgers, ice-cold Coca-Cola, and the latest *New York Times*. Could there be a more patriotic table? I had not tasted anything that cold in six months, and I greedily filled my pockets with as many cans as possible. Later, I filled my two canteens with Coke, a diabetes depository. Everything I drank for the last few months was lukewarm. You don't realize how much you miss ice when it is not around. I stuffed as many sliders in my mouth as humanly possible, Joey Chestnut style, and filled every pocket with as many sliders as they would hold. It was like a greasy, beautiful dream come true. I grabbed the Times, and the first thing I did was check the date. It was today's date! This almost brought me to tears. The cheering and flag-waving crowd that greeted us as we entered the airport was overwhelming. They had a barrier up; they were waving flags and clapping. I felt like a rock star. The girls broke the barrier and kissed us on the cheek, and I couldn't stop smiling if I tried.

We were still not home; we had one more flight to Savannah from New York and then a 30-minute drive to Fort Stewart. We boarded the next flight and were on our way. I had a lump in my throat the entire flight. We were shown the movie *Ghost*, a lovely, upbeat film he sarcastically wrote, for our flight home. The Swayzeness of the film washed over us. The film ends, and we begin our descent. We landed and got on the bus. A sergeant I've never met greets us as we load up on the bus, handing us PTO forms to fill out.

"Soldiers, I am authorized to grant you 30 days of leave right now, even if you do not have the days! You will go on leave, but I can authorize it right now. I advise you all to take the 30 days and spend them close to your families. This is what I've been doing since I got back. I don't think I've let go of my children. They are annoyed by it." He said with a wry smile.

When the bus pulls into the parking lot of the 3rd Engineer Battalion, there is a crowd waiting for us. This includes Caldwell and his family. I was happy to meet them finally; his wife and daughter were excited and clapping. Caldwell hands me a bottle of Jack

Daniels. I'm nineteen, by the way, old enough to die for my country but not old enough to have a drink. Makes no sense. Almost my entire unit was there to greet us. It was something quite moving. I'm assigned a room in the barracks; Caldwell helps me take all my stuff to my room before heading to his house to get drunk.

The next day, my leave began. I had to figure out how to withdraw my money from the bank and then how to buy a plane ticket home. This was a lot more work in the nineties. Gus finds me and takes me to the airport in his new Ford F-150. He could not afford it, but he had it.

The flight home was the longest six hours of my life. I had a direct flight. My mind kept wandering. It all felt like a dream. I didn't think I was going to make it. I could not wait to see my family and friends. It was a fantastic reunion. I could not go anywhere without people waving, patting me on the back, and thanking me for my service. I was treated like a hero, and I had a bit of impostor syndrome. I didn't feel like I had earned these accolades.

Back home, sleeping in my childhood bed and staring at my Cindy Crawford and Heather Thomas posters felt surreal. The bed was too soft; I could not get comfortable. Nothing had changed since I left; my room was exactly as I had left it. I didn't feel like the boy who slept in this room. I felt different—a stranger in my skin. Sleeping indoors felt strange, and I realized you can get used to anything. I didn't know how to talk to my friends and family about what I had experienced, so I shut my mouth and sought normalcy. I wanted pizza and burritos, and to spend time with my friends. I wanted to play mini-golf and go to the movies. I missed so much pop culture that I wanted to dive back in. I wanted to feel like my old self again, but this would never happen. He died in Iraq, and this new version of me was back in the world, taking his place. I wanted things to be simple. I had changed, and I could feel it, and they could sense it.

But, as time passed, I found myself telling my stories, trying to find humor in everything. I still need to find out if I was doing it for the listener or myself, but the stories were not amusing. My friends seemed horrified. I needed help picking up on the social cues. It was the first sign of trouble that I didn't pick up on at the time.

CHAPTER 15

The war was over for all of us. Our side lost about three hundred men, primarily to accidents. The Iraqis suffered a more significant defeat. No one knows the exact number; they lost twenty thousand to fifty thousand. For what? The ego of a madman. They looted a billion dollars and raped and pillaged the entire country. I still don't know what we accomplished.

July 1990 to April 1991 was a lost year for me. It's not a year exactly, but it's close enough. At that time, I was no longer in control of anything. I just followed orders and barely had any time to myself. It took a while for me to stop looking at the newspaper's date page. I felt free and trapped by the Army at the same time. I was also getting used to the Army way during peacetime. It was just a job now. I sat around and prepared for something I'd already done. We trained to fight in a war; that was our purpose. I fought in a battle. What was my purpose now?

Time slowed after our return—no vehicles in the motor pool. Weapons and equipment were on a ship somewhere on their way to Fort Stewart. The duffel bag I shipped with all my stuff hadn't arrived. It had a month's head start on me, and I still beat it to Georgia.

Our new captain had an insane Polish name I couldn't remember, but I could swear there was an exclamation mark. He was a triathlete and used his company to train for runs. We had nothing else to do, so our new captain tried to exercise us to death, sometimes nine miles daily. He wanted us back in shape since we had it easy while fighting a war. My knee started twitching while I was typing that sentence. My knee was/is in bad shape, and I finally got surgery a year after returning from the Gulf.

I started at 9 a.m. on a regular day in the motor pool, next door to a swamp. In summer, it would be 100 degrees with 100% humidity, permanent swamp ass. You are always wet. It was a miserable experience. Ate lunch at 11 a.m. Busy work until 5 p.m. I would then have to sit in line and wait for a payphone to free up so I could call home. My parents got me a long-distance calling card. I had that code memorized for years, much longer than I should have. We were not allowed phones in our rooms, so you had to share six payphones with eighty other soldiers. Fights would break out. It is hard to imagine nowadays not having easy access to everyone in a second.

I was bored with my work. I wouldn't even half-ass my job; I quarter-assed it. The banality of Army life takes a toll on your mental well-being. The weekends were my oasis. I still had two and a half years left to serve. I had a short-timer mentality. I had that mentality for the rest of my time in the military. Peacetime was so different. I was promoted, skipped E-2, and went right on to E-3. I eventually settled in at E-4. I had no plans to go any farther. I couldn't wait to get out. It was all busy work and training for the next conflict. I was also promoted to track driver of the 3rd squad, taking over Caldwell's position. I did this for a year before moving up once again.

Army life was very disciplined. I worked from 6 am to 5 pm every weekday. I would hang out with Caldwell after work with his family, then leave and go out with my single friends at least four times a week. I was young with a fake ID; nothing would stop me. I would be power-drinking until 3 am, get back to the barracks, pass out, and get up at 6 a.m. for PT. I would run, jog into the woods, power vomit, and then return to formation. I kept up this routine for a year after coming back from overseas. I had no parents stopping me. I was my own worst enemy. My liver wrote its congressman to no avail.

Eventually, I got to the point where I just hated how I felt every day. The first part of my day was rehydrating and nursing a daily hangover. Professional drinkers surrounded me, and I thought this was what being an adult was all about. I needed to get together and stop drinking cold turkey for a year. I did not sip alcohol for my twentieth year on this earth. When my twenty-first birthday came around, I went to the PX and bought a beer with my fake ID.

I only started drinking like that again after I got out of the Army and needed to self-medicate to sleep; otherwise, I'd be up alone with my thoughts all night, tossing and turning. I would miss the peace I felt when I was twenty.

The 3rd Engineer Battalion did not lose anyone in combat. We all made it home. We never discussed how we were affected. It was too soon to assess any real damage. Everyone seemed all right; none of my friends seemed off. If they were involved, they hid it well.

We all handle things in our own way. Even with all the vaccines and sleeping in puddles, I never got sick in the Middle East. However,

I had a nasty case of the flu when I got home from leave. It was so bad that I was quarantined in the barracks for three days.

One night, at about midnight, there was a knock on my door. I opened it and saw Gus standing in front of me, covered in blood. He had taken a razor and cut his face, cheeks, and near his eyes. He also had razor cuts through his shirt and on his chest. Some deep cuts, others not so deep, but blood everywhere. He looked like someone had dumped a bucket of blood on him.

"Hey, Joe, you wanna go to the club?" he asked cheerfully, covered in blood from head to toe. He looked like Carrie before she burned down the gym.

"Dude, what happened? Are you okay? Did you get mugged or robbed? Were you jumped into a gang? Are there native American gangs?" I replied.

I was terrified for him. Then, behind him, I saw two other friends slowly sneaking up on Gus. I decided to keep asking questions to keep him from running away. They grabbed him and forced him into their car and to the hospital. I followed them to the car; Gus was now ranting and raving. The three of us shoved him into the car. His leg got loose, and he kicked my friend in the face. We slammed the door shut, and they sped off to the hospital. I could not sleep that night; I was worried about my friend's future. The next day, I went to visit him. He was on the fourth floor in the psychiatric hold. His hands were secured to the bed. Gus was covered in bandages, bloodstains seeping through. He saw me and smiled. He was overly cheerful for someone in his situation.

"Joe, you gotta get me out of here. These people are crazy," Gus said.

"I don't think you're too far behind, man."

People were tied to their beds, some screaming, others catatonic. My heart sank—these poor people. Gus was out of the Army for a few months. It was the best thing for him. That was my first brush with PTSD, but sadly, not my last. We all react differently to trauma. Gus's trauma hit him immediately.

During the first few weeks back in the world, I had nightmares of being in the desert again. I would see the man burned into the fence, staring blankly at me. I replayed those screams over and over. I would see other moribund faces.

They spoke to me. I couldn't make out what they were saying, and I tried to help, but I couldn't. I would see rotted skeletons sitting out in the sun. They would turn and look at me. Their eyes were plucked out of their skulls. They tried to escape being pushed into the giant holes we dug.

We pushed their corpses into their final resting place. They would never be found. I would sometimes exchange places with them, and I would be the one having sand pushed over me. No one could hear me calling out as they buried me alive. I'd wake up choking and sweating.

This happened for weeks, maybe months. Slowly, these dreams stopped, maybe after a year. I never told anyone or asked if they were experiencing what I was going through. Gus was the only person I saw outwardly affected.

All I did was bury my trauma. It would slowly dig its way out over the next few years, wreaking havoc. You hear about the sights and sounds of war, but no one tells you about the smells. You can smell blood in the air. It is mixed with gunpowder and exhaust fumes. Copper on your tongue. The smell of burning hair and flesh never leaves you. It is horrific and unmistakable when you come across it. If I go to a firing range now, the second that scent of gunpowder hits my nose, I'm back in the Army, on the range, in the desert. It took me years to get over this. These memories come out at the strangest times; I wouldn't wish it on anyone.

When the equipment finally returned from Iraq, after another year, I left the 3rd squad entirely and became the track driver for the TOC (Tactical Operations Center). This wasn't based on my performance; I was the only person who knew how to type.

I was responsible for driving the captain and the XO around. I worked in the main office, the air conditioning, and only a tiny fraction in the motor pool.

Just because I was driving the captain and XO around didn't mean I was immune from work when we played war in the field. Of course, I had to help build ranges and put up razor wire, but I spent much of my time in the TOC hanging out with the captain and XO, reading USA Today, and not doing much else. It was a great job. I'm glad I took typing in high school. Plus, all my years of programming on my Commodore 64 computer.

I saw all the bureaucracy and politics firsthand. Seeing how the officers had to navigate dealing with the sergeant major and colonel, they hated it just as much as I did. I also knew when the drug tests were coming and could warn certain parties.

It was strange when new people, outsiders, came to our unit. They were hazed. We would have them jump up and down on the personnel carrier to test the shocks. The track would not move, yet they would jump up and down on it for long periods. We had them stand in different areas of the track and jump and pretend to take notes. We had the newbies go around with a hammer, hit the armor, and mark all the soft spots with chalk. There were no weak spots, and it was barely armored. We sent them on impossible errands, like finding me a left-handed crescent wrench. I was counting the days and entertaining myself by torturing the new people in the unit.

I was awarded the Kuwait Liberation Medal, Army Service Medal, Good Conduct Medal, National Defense Service Medal, Overseas Service Ribbon, and Southwest Asia Medal with two service bronze stars. I was in a total of three years and one month. I had more ribbons than most higher-ups who had never been to war. The military used this as an excellent excuse to eliminate many people, forcing early retirement. This came back to bite them later.

Towards the end of my time, our original First Sergeant Winston was replaced by my old company commander, Sgt. Slevens. He hated me, probably for leaving him hanging on that nail in Saudi Arabia. We also stole his Saudi canned soda; it was not Coke or Pepsi but some Saudi concoction, which wasn't very good. We buried it directly underneath his cot. He chastised us, knowing we had done it but could never prove it. It is still out there! Get to work, treasure hunters! I was also the Rich Little of our platoon. I could imitate his voice perfectly and used to do it over the radio. I would make outrageous statements in his voice. I would do this over the radio, and he would come looking for whoever was imitating him; no one ever gave me up.

It was great moving from working in the motor pool all day to working in the company office with the Captain and XO (Executive Officer); I do all the data entry and have my own office; it is pretty sweet. The air conditioning blows directly on me. I only had to interact with him briefly; we now worked in the same office with his new promotion, and Slevens had it out for me. This was a no-bueno situation for me. He assigned me every shit duty. I was a short-timer. I would smile and take it. I now only had four months left. I had to work three weekend duties in a row. He was deliberately fuckng with me. I walk in on a Monday and see the following weekend duty: just me and no one else. I blew my lid.

I stormed into his office and slammed the door behind me.

"Murgia, what are you doing?" Slevens asked.

"I'm not doing that duty. You can assign it. I won't show up. I will have no more duties for the next four months. I don't care if you bust me down; it isn't stopping me from getting out. I have dirt on you, and I will use it if you do not back the fuck off me. You profited off your troops, and I have the proof. I wrote down everything you marked up. You are taking twenty-five percent off the top of every sale. This is illegal." I said, and it was the truth. I'm so happy we wrote down all that evidence, and I was smart enough to keep it.

"You wait a goddamn minute." He replies and stands up in his chair.

"No, you wait and sit the fuck down; you don't fucking intimidate me, you bald fuck." I replied.

"I'm the first sergeant. I tell you when to talk. You think you can blackmail me?"

"This isn't blackmail, I don't want anything from you other than leaving me the fuck alone. You think you're untouchable because they are out of the Army. I have everyone's phone number. They may be home now, but I can contact them in a second. I speak to some guys weekly. Gus and I told them what you were doing back then. We showed them the markup on items. They knew you were robbing them, but they had no choice but to buy from you. I will get signed statements from all those troops and report to you at the Department of the Army. If I get even one more assignment that I feel is out of line, I will end your career."

He was silent and stared down the end of his nose at me. He was red and seething. I did not care. I had all of the profiteering proof of profiteering. I sent it home in my duffel in Iraq for safekeeping. I continued, "Leave me the fuck alone for the next few months. I'd appreciate it if you didn't speak to me or even look in my general direction. You'll lose that pension you're always talking about. I fucking guarantee it.

You stole my AKs, too, you piece of shit. I don't want to hear you say my fucking name again, or you are going down. So do the smart thing and leave me the fuck alone until I leave."

I then turned around and stormed out of the office. I didn't even have to make eye contact with him anymore.

I counted the days until I got out—Staff Sgt. Winston was married to an officer. She was a captain, and she was part of out-processing. She taught a class to help you write a resume and prepare you for life on the outside. This was a mandatory class. I attended with a friend who couldn't wait to get out. I remember the first thing we learned in class.

"One thing you probably haven't thought about is clothes," Captain Winston said. "Clothes make the man or woman. You've all heard this before. One thing the Army does for you is take that thought away. You don't have to think about what you wear, like when Einstein wore the same set of clothes every day. It frees you to think about other things. You wear the same clothes every day: your uniform. But being a civilian, you do. You must pick your clothes out every day before you go to work. Think about that: expending that energy every day. You must make an impression."

I raised my hand.

"Yes, specialist?"

"Are you telling me we will have to think for ourselves?" I asked.

"What's your name, soldier?"

"Murgia."

"Ah, Murgia, yes, I've been warned about you. Please don't disrupt the class. My husband told me about you. Just sit there quietly until class is over."

My captain once told me the Army should devise a manual on dealing with me. I asked "why?" a lot. Why are we doing this? Why are we marching? Why are we running? Isn't this a mechanized division? And so on. The funny thing was that nobody called me by my last name like they did every other soldier. Only Sgt. Slevens called me by my last name. I was Joe to everyone, including the Captain and XO. I was a pain to deal with, but I got the job done most of the time. I played stupid to get out of work. I couldn't stand taking orders. I wasn't cut out for the military and was only a few days away from driving home.

After the out-processing classes, I turned in all my equipment. I remember having my out-processing checklist, and there was only one thing left to do: go to the battalion and have them sign me out.

The clerk reviewed my paperwork, turned to the last page, and signed. "All right, you are officially out of the Army."

"You mean at midnight? End of day."

"No, I mean you are out right now. You are officially a civilian. Thank you for your service."

I'm free. I couldn't believe it. I almost burst into tears. I turned around, went to the barracks, loaded everything in my truck, and sped off base as soon as possible. I didn't want to give them a chance to change their minds.

Then, the weight of the world lifted off me. A new world of possibility was now open. What would I do? It didn't matter. I wanted to get home. *Once I get there, I'll figure out what's next.*

EPILOGUE

The Army did one thing very well: it gave me direction in life. I learned I would work in air-conditioned offices instead of baking in the sun doing manual labor; it is not for me. I also spent a lifetime camping in three years. I believe in service. All citizens should serve for one year, not necessarily in military service, but in some service. I would remove one year of high school and make service mandatory. You may disagree. It could be in the Peace Corps or Americorps. A year of helping others. I think this would be good for most people. If you even volunteered to wash animals and take dogs for walks at a local shelter, that should count, too. I think it is good for you and society as a whole. It was for me. It gave me structure. I still make my bed every day. If I accomplished anything all day, at least I made my bed. There is satisfaction in that.

The one thing that bothers me is how my being in Iraq affected my parents during that time. They were sick with worry. I wish they did not have to experience that pain. My dad would tell my mom every day, "This is the day we will be told he is dead." He never expressed this to me; I heard it years later. My mom would also have to calm my neighbor lady down on the regular, who had a son overseas. She thought that morbid calls would come daily, like my dad, but they never did. We both survived.

It must've been tough. It bothers me that you don't think about how your decisions affect the lives of people around you until it is too late. I wish they had never gone through it. The worry they felt is something I will never understand since I have never had children.

I signed up for community college right after getting out of the Army. I attended many community colleges over many years. I even got a bit of college credit for being in the military, but every little bit helps. I was adjusting to civilian life. I had to pick my clothes daily, so it threw me for a loop.

The PTSD started to manifest a few months after I got out of the service. In college, I noticed things about my mental and physical health that surprised me. I was easily startled; if someone dropped their books on their desk, my heart rate would go through the roof. I had insane sweating and that hollow feeling in my chest, my first panic attack. I had no idea at the time that this would be a constant companion from my thirties forward. I thought it was an anomaly, but I was wrong. I kept ignoring issues and making excuses. My knees were in constant pain, and the same was true with my ankles; I could not run like I used to. College was difficult; I had trouble concentrating and staying focused. *Why can't I focus?* I never told anyone; I was embarrassed. I finally ended up in a trade school that no longer exists and learned Computer Networking. I used the *GI Bill* to pay for it; this is why I joined, why I put my life at risk.

Even with all of these mental health issues, I was grateful that I could hold down a job. So many veterans have trouble finding and keeping employment. Most jobs don't give you the amount of responsibility you get in the Army. You are in charge of million-dollar vehicles and weapons, and then you must get a job at a car wash after you leave. It is challenging to adjust from active military to civilian life. I worked steadily from my twenties through my forties and into my fifties. *The capitalism grind.* Each job felt similar to the one before it, white, overhead fluorescent light, sucking the will to live out of you. No daylight. White walls. Each floor has a different cubicle configuration.

A new group of work friends that sometimes become real lifelong friends, some of whom you work with and never speak to again. This is the way it was in the nineties and early two thousand. I do not see an end to the grind. I think I'll work until I die, the American way.

I was able to keep my secret drinking from interrupting my work life. I needed it to sleep and turn my mind off at night. I drank about a quarter of a bottle nightly. Mornings were rough, but you power through; at least I slept terribly, but I slept just the same. I got three to four hours of sleep per night for fifteen years. I needed help, but did not know where to look or what to do. I did not consider myself a person with a substance use disorder, and it was pragmatic. I needed sleep, and alcohol got me there. I was the master of making excuses and kicking the can down the road.

Ten years after getting out, I filed my claim for all of my physical and mental health issues. I had no idea this was even an option, getting paid for the damage the military inflicted on my body and mind. Getting approved took a few years, but I got a one-hundred-percent medical. As a veteran, I have socialized medicine, and I love it. I never pay over $8 for a prescription. I wish everyone had access to this program, at least the prescription program. Socialized medicine works great for me.

I'm one of the lucky ones; I didn't end up in an institution or, worse, follow through on self-harm. I got through relatively unscathed.

I was angry at the world, but I couldn't tell you why or articulate it in any way. I spent my twenties and thirties angry and could feel myself spiraling. I couldn't do anything about it. I didn't want to live that way.

My emotions would be all over the place, manic. The highs were high, and the lows were very low. I had great friends I could talk to, but it never solved the issues. After speaking to my friends, I would be alone again with my thoughts. My thoughts scared me sometimes. I would cry uncontrollably sometimes. I wouldn't even be able to tell you why.

I had weird OCD that seemed exacerbated by my PTSD. One night before bed, I kicked off my shoes and could not sleep, knowing they were in the middle of the living room floor. I had to get up and put them in my closet, and then my brain let me sleep.

I thought about ways of offing myself so no one in my life would have to find my body. I thought about it a lot and planned it out. I would take pills out in the desert. I felt like a burden to everyone. I knew I didn't want anyone I cared about finding my body. Let the authorities deal with it. No one would miss me. I couldn't articulate the anxiety. I felt alone, but I never was. That was all in my head.

I was against brain-altering medication; this is how I looked at it at the time. I needed help and feared any medication would fundamentally change my identity, which scared me. I based this on a friend who said she didn't feel like herself after taking antidepressants. One anecdotal story sent ripples of fear through me. This put a fear of medication in me. I thought medicine would make me a walking zombie and destroy my personality. I felt I would be walking in a druggy haze. I was already experiencing this daily through a lack of sleep. No sleep kept me on edge. I never knew what might set me off. I was so wrong. I wish I could go back and talk some sense into myself, slap myself around.

At thirty-five, I told the doctor about my lack of sleep and suicidal thoughts. I could not go on like this, and he changed my life. He informed me he could give me a sleeping pill prescription, but he did not want to do this. He wanted me to try something else first: marijuana. Medical marijuana had been legal in California for fifteen years or so at that point. He wanted me to get my card and try it out first before giving me pills, something more natural to help me sleep.

My friend and I got our marijuana cards together, and I went to the shop and told the budtender about my sleeping issues. I got my prescription weed, but inhaling the smoke was the next fear I had to overcome. I'd never smoked a cigarette, much less inhaled cigarette smoke. I just had to bite the bullet. If I choked, I choked. Of course, I did not choke at all. More bullshit anxiety over nothing. That night, I slept eight hours and did not get up once. Marijuana saved my life.

Marijuana helped my sleep issues, but other strange problems started cropping up. I have weird heart palpitations and suddenly cannot catch my breath. They began as minimal, minor episodes, but they happened more frequently. I was driving and had to pull over. I was waking up covered in sweat and hyperventilating in a mid-panic attack. I missed work. I thought I was dying. My heart was out of rhythm. I could feel it. It always started with a hollow feeling in my chest. I couldn't take it anymore; I needed more help. My father had heart issues. I thought I inherited them, and quite frankly, I was afraid to go to the doctor and get this diagnosis; I'd rather not know.

I went to the VA and started talking to doctors. It was a long process, with many emergency room visits—too many to count. We ruled out heart issues. My anxiety about having a panic attack would induce a panic attack. I have no idea what set them off. It was a vicious circle, and I had had enough. The panic attacks started once a month, then once every couple of weeks, and then finally daily.

This happened so often that I gave my fear and anxiety a name: "No Joe." *No Joe is a liar. No Joe tells me I'm fine. I'm not okay. No Joe tells me to stay inside all day. Retreating into yourself is OK. It isn't acceptable. No, Joe doesn't like exercise; dopamine hit. No, Joe, she doesn't want to talk to you. No, Joe, no one cares what you think.*

No, Joe, they especially don't care how you feel. No, Joe, why bother? You're just going to get rejected. No, Joe, don't put your dick in there!

Sometimes, *No Joe* was right. Mostly he was full of shit and prevented me from living. I didn't know I had a *Yes Joe*, always drowned out by fear and anxiety. *Yes Joe* got quite a bit louder as I worked on myself; I had to tamp that fear down. It is like those cartoons with the angels and devils on your shoulders. There was no angel, just the devil, shouting at me, drowning out any confidence I might have had. Fear took over.

No Joe was exhausting, and I got to the point where I started having evil, intrusive thoughts, something in the way of making me feel any level of happiness. I was surrounded by great friends who were supportive and there for me, but even that love couldn't stop *No Joe* from ruining my life. I would make a doctor's appointment, fear would take over, and I would be a no-show or cancel.

This cycle continued for about a year: set the appointment, cancel, repeat. *No Joe* did not want to be fixed or dulled; the fear got louder after another massive panic attack at work, which caused me to pass out. My friend found me lying in a bush outside our office—no idea how I got there. I remember going outside, and that is it. It was never good to be found lying in bushes.

The ambulance took me to the VA hospital about three miles away. It was a $1200 ambulance ride, and they didn't touch me in the back of the ambulance. You have to love the American healthcare system!

I knew I had to get a hold of my anxiety in addition to taking the medication. I knew what to look for in the beginning stages of a panic attack. I learned breathing exercises and how to relax, identify what was triggering me, and remove myself from that situation. It took me years to figure out that I was giving *No Joe* all his power by giving in. Nothing worked; drugs stopped working, drinking stopped working, and the love of my friends and family stopped working. I hit a wall. I needed help. I was afraid to ask for help. *I'm taking No Joe's power away.*

My anxieties were unfounded, as always. I started the medication the afternoon after my hospital visit and have not had one panic attack since. Well, that is not precisely true; one time, I woke up in the middle of the night in a full-on panic attack. I was able to identify what was happening and willed it away. I was so proud of myself. I pushed back on the fear and won.

The doctor gave me an examination, and he concluded that it was anxiety; this was the conclusion of all my unplanned hospital visits. I had no other health issues.

He handed me a bottle of antidepressants. I was still wary of the pills, but this event pushed me to finally commit to taking them. I didn't want to make waking up passed out in the bushes a regular occurrence.

What does Tom Petty say in the song "Crawling Back to You"?

I'm so tired of being tired.
Sure as night will follow day
Most things I worry about
It never happens anyway.

This was my whole life, worrying about something that would never happen. You feel like an asshole. One time, when I was at the hospital after a massive panic attack, I spoke to a retired colonel who told me that just being in a combat environment changes your brain waves forever; you are no longer like everyone else, and there is no going back. This is why veterans struggle to integrate back into society and connect with others. You are simply out of sync with others. The experience of combat can be highly stressful. It can trigger a range of neurological responses designed to help soldiers cope with the danger and uncertainty of their situation. I certainly felt this in real time. Fight or flight, we ran towards danger, not away. It is not natural to run toward danger; every fiber of your being is fighting this response. This fucks you up, and it takes years to manifest. Also, running from danger, I guess, could be considered another nugget of wisdom. This is a good one to remember in a book with so little wisdom.

Desert Shield/Storm may not have affected the world and culture as much as it affected me. It was a precursor to the world we live in now. We lived during a long, peaceful time, and one asshole, Sadaam Hussein, fucked all of that up. The next world-shattering event happened eight years later. I can't even remember what a peaceful world felt like. I understand there is always a military conflict somewhere, even during the peace and prosperity of the nineties, but it was typically buried on page six of the newspaper.

My time in the military was the best time of my life in many ways. I was away from home, with no rules and a definite sense of finality if I made the wrong choice, which I did several times. I never felt that alive before or since. I am still in contact with many of the men I served with, some of whom are friends and others are not. I lost contact entirely with a few. I searched for years. Even with the advent of social media, these people are either not on social media or, sadly, no longer on Earth. I've never stopped looking, even contacting the Department of Defense, but still no luck.

When I tell people I served in Iraq, they think I'm speaking about the war in 2003 after 9/11. I then explain that I mean the other battle with Iraq in 1991. I'm usually met with, "Oh, yeah. I remember that." The war I fought in seems to be filed away in people's minds next to the forgotten bombing of the World Trade Center in 1993. They tried to take the buildings down back then, too. 9/11 is the reason the war I fought in became so culturally insignificant. It makes sense that since that event shook the world to its core, Desert Storm seemingly only affected the soldiers who fought in it. Desert Storm had no over-arching cultural or political narrative. This lack of a straightforward narrative meant the war could not capture the public's imagination or inspire significant artistic expressions. Moreover, the war was fought in a relatively confined and isolated region, which limited its impact on the global stage.

I miss the nineties. I never thought about politics or being narcissistic enough to believe the world would end in my lifetime. I see that as a real possibility now. I try to be optimistic, but pessimism keeps landing uppercuts on my psyche. All this information flooding our senses daily isn't healthy or good for us. We are too busy dealing with the latest outrage in the twenty-four-hour news cycle. We move on quickly from tragedy because something worse is happening and swallowing up the next news cycle.

I blame this on social media, where bad news travels like wildfire across the internet, whether the news is real or not. We've typically moved on to something else when the news we were outraged over yesterday is debunked. The world was not run on conspiracy theories in the nineties. Half the population seems to be wearing tinfoil hats nowadays. This reminds me I must resize mine; my giant Irish head keeps expanding.

Desert Storm was four days long (six days for us on the ground). It wasn't altering society like Vietnam did. It was a jingoistic time in America. We were riding high as a country, and not even a minor war would stand in the way of all the good things happening in America.

Everyone I served in Iraq with left the military before me. One by one, they went, and new troops eventually replaced them. We treated the new troops poorly; they were outsiders and didn't serve in the war with us, so they were not seen as equals. Eventually, everyone I fought side by side with was gone. The Army used the war to retire many people early, which backfired a few years later since they let too many people out. This would affect me a couple of years later when I tried to get an education drop to get out of the Army six months early. The rest did their time and got out while the gettin' was good.

Sgt. Martinez died years later after another battle—cancer. We spoke often. I talked to him a few weeks before his death and reminisced about old times. It was heartbreaking. He did not sound well, the spark taken out of his voice. He was a great man. Kind. Cancer sucks. He was with me during two of my worst experiences, and without him there, I probably wouldn't have handled it as well as I did. He was a calming force. I miss him. He visits me in my dreams sometimes; he seems happy.

I lost touch with Sgt. Steward. He was promoted and transferred to another unit. I never got to thank him for all that he did for me. I spent many years looking for him after I got out of the Army, with no luck. He kept me close and kept me alive. He felt tremendous responsibility for the lives of the people in his command. You could feel it. I owe him the world.

I remained good friends with Caldwell until he left the Army several years later. Bunk was around, exited the Army, and returned home to Ohio. I spent years looking for him after I got out. I am still looking for him. Johnson remained in the squad until I switched jobs a year later. Corporal Peel also moved to another squad, and then he was gone. These people enter and leave your life, but the memories remain forever. Peel filmed a ton of footage; I often think about where it could be. I would love to watch it.

Our squad had such a tight bond. War buddies are not a bond easily broken. Unless you support insurrectionists, then the choice is easy for me. You are removed from my life. I guess some forgot what they fought for. I did not. I defended democracy and the Constitution, as well as cheaper gas. It breaks my heart to cut people out of my life, but if you are on the wrong side of history, it is an easy decision, just painful. It is a shame we have become this divided. The government loves us fighting with each other while they steal from us and give more tax breaks to the rich. It is us against the government, not left versus right.

I will always be connected to those men I served with. We didn't always agree or see eye to eye, but we were there for each other. When we reconnect, it is like no time has passed. We are right back there again. When I first met my unit, I thought, "I'm not surviving this." They were joking around and never seemed to take the situation seriously.

I realize now that it was how they were dealing with the problem. Gallows humor is necessary. When the war started, and shit got real, they all tightened up and worked together. We were a unit. My first impression couldn't have been more wrong. I learned to respect these men, not all of them, but most of them.

Not a day goes by that I don't think about my time in the war, even more so since I started writing this book and speaking to the Army buddies I served with. I don't see ghosts that much anymore. I don't think they will completely disappear, but I'm at peace now.

What was the point of Desert Storm? To help Kuwait? Kuwait was barely mentioned after we got back home. I've never met someone from Kuwait. I felt like I was there for America and nothing more. What is a win in the Middle East? What does victory in the Middle East look like? I have no idea. I've thought about it for a long time. There doesn't seem to be an answer.

We seem to be stuck in the same cycle. Some disgruntled group attacks us, and we retaliate with no clear objectives. Hell, we bomb countries that had nothing to do with the attack in the first place. Our first thought is just to bomb any brown people who live in that general vicinity. We keep making terrorists by killing their families, and they grow up wanting revenge.

Add in the fact that there is no economic opportunity in that part of the world. You don't get a job at McDonald's at sixteen or a pizza delivery job. Those jobs don't exist. They barely have an infrastructure in that part of the world. In America, these jobs are a rite of passage; no such thing exists in the Middle East.

This is why young men are so easy to recruit in that part of the world. There is no better option. Thanks to robots, drones, and AI, those jobs will also disappear in America. Will our future kids be radicalized when they also have no job opportunities? How do we fix it? Nothing will change unless jobs and opportunities can be dropped from a plane or drone.

After 9/11, America continues to throw money at a problem that will never be solved: the same issue, different decade. I wish I had the answers, but I believe the answer is that *there is no answer*.

War props up our economy and advances technology. War is capitalism. People must die for the economy to keep spinning along. This is a sad fact. Keep the poor stagnant and make the rich richer. We need cannon fodder, so not everyone can be rich. This has been going on since humans could stand upright. It will never end. The young are dying for the old in power. It appears that killer flying AI robots will be risking their microchips. Is it better? Yes, but I've seen this movie; I know how it ends.

I think about my idealism and blind adoration of our flag in the early nineties. *We can do no wrong, these colors don't run, blah blah blah*. We must find a way to break the cycle. The Middle Eastern countries have been fighting with each other for centuries. This will never change, and there will never be peace. *Why don't we stay out of it?*

I know—it's very idealistic and unrealistic. I now have a mantra and do my best to live by it. I try not to think about the future or past, but to live in the moment. This isn't easy. It takes practice, and I don't always succeed. I'm guilty of wallowing in nostalgia, but it is like a warm blanket.

If I catch myself not living in the moment and focused on some nonsense, I stop and tell myself, "The future hasn't happened, and there is nothing you can do to change the past. It happened; there is nothing you can do to change it. Focus on now."

This works for me. It may not work for everyone, but I've reached a good place. I've dealt with my past, and you can too. It helps keep me on a healthy mental path. I've spent 20 years repairing the damage from the Army and war. It was an uphill battle; anxiety had the high ground. Anxiety kept me from living my best life. The medicine works; I'm not a zombie. I still feel like me. I don't let *No Joe* control me anymore. I can't hear him anymore. I pushed him way down. *Fuck that guy.*

In November of 1993, my time was up. I turned in all my equipment and went to all the out-processing classes. I went to the battalion, showed them my completed paperwork, and signed my name. I was officially out of the Army. The world's weight lifted off my shoulders, and I cried in the barracks and got myself together before I threw all of my stuff in the back of my Chevy S10 and hit the road home to California. Exiting through the gates of Fort Stewart that last time, I felt like it was the first day of the rest of my life—the end of a chapter. I felt reborn as I hit Interstate 10 West with a calm wind in my hair. A new chapter is beginning—an exciting adventure into the unknown. I couldn't have felt more grateful.

Book Credits

Memories From a Culturally Insignificant War

Written and Edited By Joseph Murgia

Published by THC Publishing

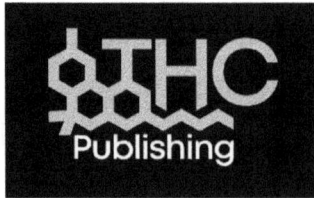

Co-Editors

Mary Lou Heater & Clara Abigall

www.ingramcontent.com/pod-product-compliance
Lightning Source LLC
Chambersburg PA
CBHW071216090426
42736CB00014B/2852